Introduction to Electro-Optical Imaging and Tracking Systems

The Artech House Optoelectronics Library

Brian Culshaw, Alan Rogers, and Henry Taylor, *Series Editors*

Acousto-optic Signal Processing: Fundamentals and Applications, Pankaj Das

Amorphous and Microcrystalline Semiconductor Devices, Optoelectronic Devices, Jerzy Kanicki, ed.

Electro-Optical Systems Performance Modeling, Gary Waldman and John Wootton

Field Theory of Acousto-Optic Signal Processing Devices, Craig Scott

Highly Coherent Semiconductor Lasers, Motoichi Ohtsu

Introduction to Electro-Optical Imaging and Tracking Systems, Khalil Seyrafi and S. A. Hovanessian

Introduction to Glass Integrated Optics, S. Iraj Najafi

Optical Control of Microwave Devices, Rainee N. Simons

Optical Fiber Sensors, Volume I: Principles and Components, John Dakin and Brian Culshaw, eds.

Optical Fiber Sensors, Volume II: Systems and Applicatons, Brian Culshaw and John Dakin, eds.

Optical Network Theory, Yitzhak Weissman

Principles of Modern Optical Systems, Volume I, I. Andonovic and D. Uttamchandani, eds.

Principles of Modern Optical Systems, Volume II, I. Andonovic and D. Uttamchandani, eds.

Reliability and Degradation of LEDs and Semiconductor Lasers, Mitsuo Fukuda

Introduction to Electro-Optical Imaging and Tracking Systems

Khalil Seyrafi
S. A. Hovanessian

Artech House
Boston • London

Library of Congress Cataloging-in-Publication Data

Seyrafi,Khalil.
Introduction to electro-optical imaging and tracking systems / Khalil Seyrafi, S.A. Hovanessian.
p. cm.
Includes bibliographical references and index.
ISBN 0-89006-672-8
1. Electrooptical devices. 2. Imaging systems. 3. Tracking radar. I. Hovanessian, Shahen A., 1931- . II. Title
TA1750.S484 1993 92-32242
621.381'045—DC20 CIP

British Library Cataloguing in Publication Data

Seyrafi, Khalil

Introduction to Electro-optical Imaging and Tracking Systems
I. Title II. Hovanessian, S. A.
621.36

ISBN 0-89006-672-8

685 Canton Street
Norwood, MA 02062

International Standard Book Number: 0-89006-672-8
Library of Congress Catalog Card Number: 92-32242

10 9 8 7 6 5 4 3 2

Khalil, when you left us, I felt that I needed to publish your book. I remember many late nights when you were sitting at your chair, deep in your thoughts, working on this book. I remember when I complained about how much time you were devoting to the book that you said to me through your beautiful eyes "Mahboob, if you only knew the material in this book, time by itself would not mean anything to you," meaning that time is a relative concept.

I am sorry that you are not here to see the results of your work. I am deeply grateful to our dear old friend Shahen for putting the book together. I know that you always thought highly of Shahen and I am glad that he assisted with this book. I know that there is much in the book that you wanted to share.

Love,
Mahboob

Contents

Preface

Electro-optical systems are devices that operate in the visible or infrared regions of the electromagnetic spectrum and are used for detection, acquisition, and imaging of objects of interest. The design and performance evaluation of these systems requires a knowledge of several technical fields, including optics, detectors, noise theory, atmospheric effects, and signal processing. Although excellent texts are available in each one of these fields, textbooks extracting the relevant parts of this knowledge as applied to the design and performance evaluation of electro-optical systems are very few. The present text attempts to fill this void and integrate in one reference the technical information required for the engineering design and evaluation of these systems.

The book starts with a discussion of historical developments of electro-optical systems and their use throughout the years in both commercial and military applications. Electro-optical system design and performance equations are derived and practical examples of the utilization of these systems are given. Spaceborne electro-optical system applications as exemplified by visible and thematic mappers used to obtain images of Earth are also included.

The book consists of eight chapters. Chapter 1 gives the historical background of electro-optical system development, starting from the observation of Leonardo da Vinci that light might be a kind of a wave to current astronomical and spaceborne applications of these systems. Chapter 2 gives a discussion of electromagnetic spectrum and optical radiation as the source of energy for electro-optical detection systems. Chapter 3 discusses the atmospheric effects on the transmission of optical waveforms and the methods for quantitatively including these effects in system design. Chapter 4 includes a discussion of spectral, spatial, and temporal variations of optical and infrared background. Since targets of interest are invariably detected in the presence of background radiation, the discussion of this chapter is of prime importance. Chapter 5 gives the classical detection theory in electro-optical system analysis. This theory is primarily based on detection of sinusoidal signals in Gaussian background noise. The advent of digital computers and the resulting computational capabilities may be used to simulate various background conditions other than Gaussian, if the available data warrant it. In these cases the classical theory may be considered in the preliminary system design analysis. Chapter 6 gives the equations used in electro-optical system design and performance calculations. These equations relate collected signal energy to competing background noise or clutter for detection. The applications of the contents of the previous chapters are discussed in Chapter 7 where ground-based, airborne, and spaceborne applications of electro-optical systems are described through numerical examples.

Most of the material in the aforementioned chapters deal with passive systems where the signal received by the electro-optical system does not originate from the system. Laser systems can be considered as active electro-optical systems that generate their own power, and the received energy is the reflection of this power from the target. Chapter 8 gives a discussion of laser systems together with several applications of these systems.

The original manuscript of this book was started by Dr. Seyrafi. After his untimely death, the manuscript was updated and more recent developments were added by myself.

Acknowledgements

The authors wish to acknowledge the support of the Aerospace Corporation and in particular Dr. Alfred N. Sorensen in the preparation of this textbook. The following individuals were also instrumental in reviewing the text and making valuable comments, Drs. Robert J. Chambers, Terrance S. Lomheim, Frederick S. Simmons, and Mr. John R. Parsons and Peter Hahn. The assistance of Mrs. Milrae Kruse Seaman in typing the manuscript is also acknowledged.

S. A. Hovanessian
Los Angeles, California
January 15, 1992

Chapter 1
Historical Development

1.1 INTRODUCTION

A fundamental change in the use of optics in the second half of the twentieth century was initiated by the introduction and integration of optical telescopes or cameras with solid-state detectors and electronics. This launched the new era of electro-optics.

An electro-optical (EO) sensor basically consists of (1) optics, a device that collects and transmits a light beam; (2) a detector, which converts optical radiation to an electrical signal; and (3) a series of signal processing electronics (see Figure 1.1). Once the optical signal is focused on the detectors, an electrical signal is produced and fed to a preamplifier and a filter. After filtering and further amplification, the signal is transmitted for more processing to a local receiver (or to an antenna for transmission to a distant receiver). Then, in the receiver, the original image of the scene is reconstructed. A television camera, with its transmitter and receiver, exemplifies a complete set of an EO system.

Whereas the history of optics goes back about 5,000 years, the history of electricity and electronics is only 200 years old. The earliest optical device, prompted by the human desire to see one's own image, was the mirror. Bronze or copper mirrors were made as early as 2000 B.C.; an almost perfect one was recently discovered in the pyramid of Sesostris II (1900 B.C.) in the Nile Valley. Optics made few advances through the classic Greek period, the Roman Empire, the Islamic period, and the Renaissance. The major accomplishment came in the period from the seventeenth through the nineteenth centuries when Galileo, Newton, Huygens, Fresnel, and Fraunhofer, among others, advanced the science of optics and provided a completely new generation of optical instruments.

Electricity and electronic industries, however, are relatively new. Although amber and the compass were known in the ancient world to have electrical and magnetic properties, no significant progress was made until the latter part of the eighteenth century. Fundamental discoveries and innovations in these fields in the period 1775–1865 culminated in the formulation of the classical equations of electromagnetism.

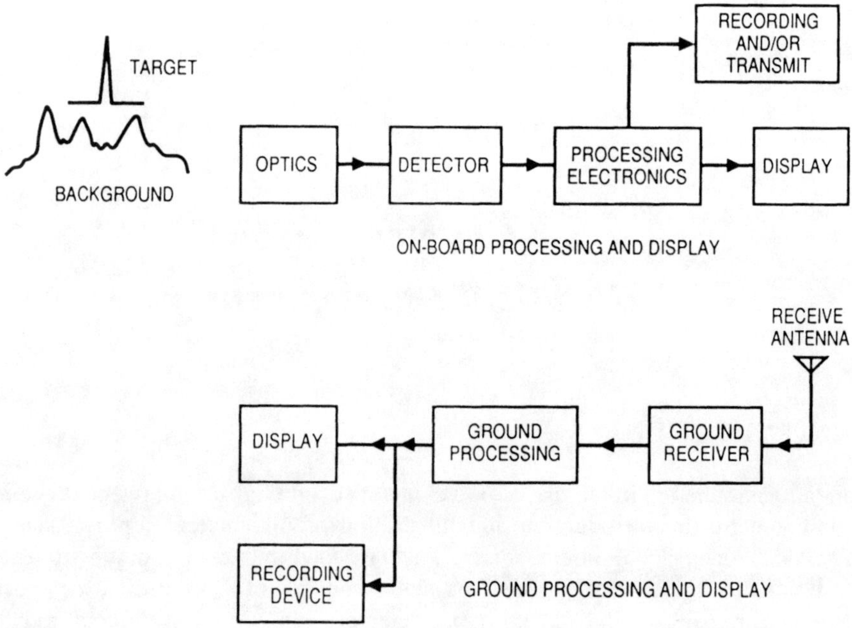

Figure 1.1 A typical block diagram of an electro-optical system.

The work of Coulomb, Orsted, Ampère, Ohm, Gauss, and Faraday paved the way for integration and formulation of the electromagnetic wave theory by Maxwell. In particular, the discovery by Weber and Kohlrausch of the relationship between the electric permittivity and magnetic permeability constants with c, the speed of light, led Maxwell to infer that light is also an electromagnetic wave. Maxwell's electromagnetic wave theory, which integrated electricity and magnetism with optics, later was verified by Hertz, who detected and generated radio waves in the latter part of the nineteenth century.

In the early twentieth century, the discovery of the photoelectric effect by Einstein made the crucial transformation of photon and electron and put the interaction of light with matter on a solid theoretical basis. This advance later resulted in the development of solid-state detectors and EO sensors.

The evolution of optics through the ages has been impressive. From the mythical beliefs of the ancient Summerians and Greeks to the discovery of the cosmos in the early seventeenth century, from the determination of temperature and composition of distant stars in the nineteenth century to the unlocking of the mystery of the atom in the twentieth, optics has influenced science, religion, and philosophy.

The merging of optics with electricity and magnetism, an accidental result of nineteenth-century experiments, set the stage for Einstein to formulate the theory of relativity. In the early twentieth century the velocity of light was established as the limiting speed of any matter in the universe. Before midcentury, intensive measurement of the optical spectra of natural elements had opened the atomic world to exploration. Perhaps the most spectacular contribution of optics is attributable to this discovery of the atomic world: the development of the atomic theory in the early twentieth century. The optical signatures of basic elements, such as hydrogen and helium, were used in continuing efforts to confirm or modify the new theory. Through the integration of optics with electronics and their application in space probes, optics has played an important role in recent years in the achievement of outstanding scientific advances.

The advent of photography in the early nineteenth century and motion pictures in the latter part of that century, both related to optics, transformed the social and cultural habits of humankind. The merging of optics and electricity in the twentieth century, giving birth to the field of electro-optics, led to the development of television. Following on a long series of unrelated innovations beginning in 1820, the invention of television in the mid-1920s not only enhanced motion picture production but created an entirely new medium of entertainment and education.

Ironically, the merging of optics with electricity and magnetism (from the work of Faraday and Maxwell), the discovery of the photoelectric effect (by Einstein), and finally, the association of optics with atomic particles manifested by quantum mechanics revealed light itself to be more mysterious than ever.

1.2 EARLY ACTIVITY

From the beginning of civilization to the early seventeenth century, progress in optics and the understanding of light was extremely slow. During this period the human eye was the only optical device to explore and be explored. The meager discoveries of the period were limited to evaluations and assessments of visual perception of the environment. Light was considered the source of and the reason for vision. Seen as a simple element related to fire, one of the four elements believed to compose the Earth, light was observed to be abundant during the day and virtually nonexistent at night. Human activities were performed during the diurnal period of light, and the human rest period coincided with the absence of light.

One of the dominant scientific figures at the end of the prescientific period, was Leonardo da Vinci (1452–1519). From the similarity between sound echoes and optical reflections he inferred, and was the first to state, that light might be a kind of wave. First analyzing wave motion in water, da Vinci moved from water to sound and then to light: "Just as a stone thrown into water becomes the center and cause of various circles, sound spreads in circles in the air. Thus every body placed in the luminous air spreads out in circles and fills the surrounding space with infinite likenesses of itself and appears all in all and all in every part." Da Vinci deduced that the eye operated as a camera obscura, which consisted of a pinhole in a wall admitting light into a darkened room; an inverted image of the outside scene appeared on the opposite wall. For centuries, this device was used to observe solar eclipses.

The period from the seventeenth through the nineteenth centuries was marked by tumultuous scientific activity. Optical innovations led to significant advances in physical science and astronomy and helped to restructure human social and cultural behavior. The period began with the invention of the telescope, which revealed new truths about the cosmos and negated ancient beliefs. Many phenomena formerly seen as mysteries were explained.

A distinguishing characteristic of this period was its novel approach to establishing scientific facts by experimentation, application of mathematics, and formulation of observed results. Applied to optics, these new methods revealed that light had extraordinary properties, never before realized. For example, it became clear that color composition was a capricious expression of light, as exemplified by the magnificent color spectrum produced when white light was passed through prisms and through glass (refraction), systematic shading of its components from red through ultraviolet, the beautiful reflection of light by raindrops, and the rainbow. Such phenomena were enthusiastically investigated and analyzed throughout the period.

Two new theories of light were developed: the particle theory and the wave theory, which competed fiercely with each other. Both theories helped to remove the 2,000-year-old visual ray theory from further consideration.

The particle theory of light dominated the latter part of the seventeenth century and the entire eighteenth century, and the wave theory started to gather momentum only in the early nineteenth century. By the second half of that century, however, it had generally been accepted by the scientific community as the correct theory.

The golden age of electricity and magnetism was concurrent with the development of optics. Experimental investigations from 1775 to 1865 in electricity and magnetism brought about significant discoveries, but the crucial relationship between electricity and magnetism was not discovered until 1850. This relationship, together with experimental results in optics, led to a turning point in science: the significant discovery that light is an electromagnetic wave. The period concluded with disclosure of the celebrated classical equations of electromagnetism. Shortly afterward, radio waves were generated and tested, confirming the wave nature of electromagnetism.

In the nineteenth century researchers discovered that each natural element expresses itself distinctively on the optical spectrum. By applying this technique in determining the optical signatures of the elements on Earth, scientists were able to establish the composition and temperature of stars. Once techniques for measuring the speed of light had been developed and their accuracy ascertained, a snowball effect followed.

The nineteenth century also witnessed the creation of the amazing world of photography and motion pictures. Photography started the process by recording and preserving a fixed or a moving permanent optical signature of human and other objects. With the perfection of techniques for the camera, phenomena invisible to the naked eye—such as the trajectory of a bullet, the movement of splashing water droplets, or the motion of a flower's petals as it starts to bloom—became visible. Furthermore, the invention of the microscope opened the new world of microorganisms, hitherto undetectable by the human eye.

1.3 INVENTION OF THE TELESCOPE AND THE MICROSCOPE

The period from the seventeenth century to the beginning of the twentieth century was one of the most flourishing eras in the history of science. It marked the introduction of a scientific approach to the field of optics, which led to the invention of the telescope and the microscope and hence to the discovery of other planets and their moons and the discovery of microorganisms. The period was also marked by a confrontation between scientific findings and church dogma.

It is unclear as to who actually invented the telescope, but records in the archives at the Hague show that on October 1, 1608, Hans Lippershey (1570–1619), a Dutchman, applied for a patent on the device. Several years later Galileo Galilei (1564–1642), an Italian mathematician, astronomer, and physicist, having learned about the patent, constructed the first astronomical telescope with a magnification power of 32. Using his telescope, Galileo, in a series of discoveries, found Jupiter's moons and Saturn's rings. He also observed that the moon's surface is irregular, not smooth, that the Milky Way is a mass of countless stars gathered together in clusters, and that Jupiter has four satellites or moons. He found that the shape and the form of these celestial bodies are similar to those of Earth. He also observed the phases of Venus, and he found "spots" on the sun. Galileo's detection of sunspots helped him to prove that Earth revolves around the sun. His observations confirmed the Copernican model of the universe, which holds that the moon, Jupiter and its four satellites, and the other planets, as well as Earth, revolve around the sun.

Galileo's celestial findings were followed by the work of a great German astronomer, Johannes Kepler (1571–1630). Kepler, who devised a model of the rotational movement of planets, was the first to maintain that their orbits around the sun were elliptical rather than circular. He was also the first to notice that the eye lens focuses an image on the retina; he proposed that a blurred image results from improper focusing. Kepler further analyzed the operation of thin-lens optical systems, including the Keplerian telescope, which uses a positive eyepiece, and the Galilean telescope, which uses a negative eyepiece. In *Dioptics*, Kepler discussed the total internal reflection and the small-angle approximation to the refraction laws.

In accordance with these experimental investigations, the mathematical formulation of optical laws initiated by Kepler began to take a more serious form. In

1621 the Dutch mathematician Willebrod Snell (1580–1626) formulated the law of refraction, a concept originated by Ptolemy in the second century A.D. Snell derived the relationship between the incident angle (the angle the incident ray makes with a plane normal to the interface) and the refracted angle (the angle the transmitted ray makes with a plane normal to the interface). He related the two angles by a quantity he called the *index of refraction of the medium*, the ratio between the speed of light in a vacuum and its speed through that medium. This relationship facilitated the mathematical analysis of the interaction of optical rays with lenses and thus spurred the study of optics.

Snell's equation relating the incident and refracted rays is

$$n_1 \sin\theta_1 = n_2 \sin\theta_2 \tag{1.1}$$

where

n_1 is the index of refraction in the first medium,
n_2 is the index of refraction in the second medium,
θ_1 is the angle of the incident ray with the normal,
θ_2 is the angle of the refracted ray with the normal.

In 1637 René Descartes (1596–1650) reasoned from metaphysical precepts that light is a "pressure" (wave) transmitted through the omnipresent "ether." He assumed that ether is a perfectly elastic medium permeating the entire universe through which no energy is lost in transmission. Following Snell, Descartes formulated the modern version of the theory of refraction. He suggested that light travels faster in dense media than through air (a hypothesis later proven incorrect). His work, together with Snell's, formed the foundation of geometrical optics and the basis for analysis of the astronomical telescope. For the first time the location of the image of an object could be predicted by ray tracing through air and glass.

The French mathematician Pierre de Fermat (1601–1665) was the first to suggest that light slows down when passing through dense media. In 1657 he derived a law of refraction different from Snell's. In formulating the law, de Fermat assumed a variable "resistance to light" in different media, presenting his "principle of least time," which was similar to the idea originally put forward by Hero in 1500. De Fermat suggested that among all the paths between two points, light travels the path that takes the shortest time. He verified this principle by using rays reflected or refracted from plane surfaces. Later, this principle was shown not to be true for rays reflected from curved surfaces.

The study of phenomena produced by the interference of light rays began with the Italian Jesuit Francesco Grimaldi (1618–1663) in 1660. Grimaldi made the first significant study of the diffraction of light, a phenomenon whereby light rays apparently bend around obstacles in their path. He observed that the edge of the shadow thrown by a sharply defined object is not correspondingly distinct. Rather, it is surrounded by alternating light and dark bands, decreasing in intensity toward the shadow. This discovery, indicating that the propagation of light is not exactly rectilinear, suggested the existence of a wave motion for light.

In 1665 two British physicists, Robert Boyle (1627–1691) and Robert Hooke (1635–1703), observed a similar effect in the structural arrangement of color rings in thin films, such as those formed by soap or oil floating on water. Hooke suggested that the reflected rays from the front and back surfaces of the thin film cause the colored interference pattern. (The famous Newton's rings, which Newton observed by looking through a convex lens laid upon a flat piece of glass, are also caused by the interference effect.) The periodic nature of these phenomena led Hooke to hypothesize that light vibrates through a homogeneous medium such as ether, rapidly generating a new sphere of light at each vibration. This hypothesis was the beginning of the new wave theory of light.

In the middle of the fifteenth century a Dutch student of natural history, Anton van Leeuwenhoek (1632–1723), developed a low-powered glass magnifier for viewing insects. In *Micrographia*, an essay on microscopes, Robert Hooke recorded further early observations of microscopic reality, including studies and illustrations of the crystal structure of snowflakes. Through his work in microscopic fossils he became one of the first proponents of a theory of evolution.

In 1830 the English surgeon Joseph Lister (1786–1869) discovered that when two separate achromatic lenses are combined to act as a single lens, the image rendered is free of any distortion caused by the curvature of the lens.

1.4 PARTICLE THEORY AND WAVE THEORY OF LIGHT

The second half of the seventeenth century was marked by the accomplishments of two of the most impressive figures in the field of optics, Isaac Newton of England (1642–1727) and Christian Huygens of Holland (1629–1665). During this period Newton propounded his revolutionary corpuscular theory of light and color, and Huygens solidified the theory of light propagation on the basis of the wave theory.

Newton demonstrated the fundamental nature of white light by showing that a diffracted light beam from a prism disperses into components of well-defined bands of color. By showing that each color has a unique index of refraction, he proved that the effect is a fundamental property of light, not of crystal.

Newton's corpuscular theory is based on the idea that light is a concrete substance composed of particles, with different sizes correlated with different colors. Together, the particles blend to form light. Red corresponds to the lowest frequency and violet to the highest. Although the results of his experiment could be explained on the basis of the wave theory, Newton chose instead to base them on the corpuscular theory. The main reason for his decision was that the phenomenon of the rectilinear propagation of light could not be explained by the wave theory.

Another of Newton's contributions to optics was his invention of the reflective telescope. The telescopes built by Galileo and others were refractors, which contained large lenses and had chromatic aberrations. Erroneously believing that chromatic aberration was incorrigible, in 1671 Newton focused his attention on using mirrors to avoid this defect. He used two concave mirrors and a flat mirror in his telescopes. The mirrors were smaller than those of the Galilean refractor pieces. Newton's first reflecting telescope was 6 in long and had a magnification of 40. Later, it was shown how chromatic aberrations in a refractive telescope (a Galilean type) could be corrected.

Newton's early studies culminated in 1704 in his comprehensive mathematical treatment of the major properties of refraction, diffraction, interference, polarization, and dispersion (variation of index of refraction with wavelength) in his book *Opticks*. In this plenary work Newton did much to establish and advance the field of optics. Unfortunately, his book inhibited further progress because of his bias toward the discrete particle or "corpuscular" theory of light. Newton's reputation, gained so deservedly through his triumphs in formulating calculus and mechanics, prevented more accurate theories of optics from being taken seriously. His elegantly constructed but fundamentally inaccurate picture of the nature of light dominated serious study for an entire century, until the double-pinhole experiment of Thomas Young in 1801.

Newton is also credited with the development of spectroscopy, the science of the analysis of the spectrum of glowing gas. He observed that a gas glows when it is heated or exposed to an electrical discharge. The radiation spectrum of the gas contains intense lines that are unmistakable characteristics of that gas. Refinements in spectroscopy and the analysis of radiation line spectra led to the development of the atomic theory in the first half of the twentieth century.

In 1675 Ole Romer (1644–1710), through observation of the eclipses of Jupiter's moons, determined that the speed of light is finite, not infinite. He estimated it as 3×10^{10} cm per second.

In 1678 Huygens strengthened the wave theory of light mathematically. He proposed that "if a point of the ether is illuminated by a ray of light, it produces a disturbance, or 'wavelet,' that fans out in all directions, forming a spherical wave. The 'envelope' created by the common convolute of all these wavelets becomes a new 'wave front' of the light ray." From the geometric construction of this principle, Huygens derived the laws of reflection and refraction. He concluded that the index of refraction of any medium is the ratio of the speed of light in a vacuum to its speed in that medium.

Newton's particle theory and Huygen's wave theory remained in strong competition for scientific acceptance until the mid-nineteenth century, when James Maxwell combined light with electromagnetic waves and generated his celebrated electromagnetic equations.

In 1801 British physicist and physician Thomas Young (1773–1829), by studying the power of accommodation, discovered how the eye's lens changes shape during focusing. Turning his attention to light, he devised an ingenious experiment that reinforced Huygen's wave theory. Young passed light through two pinholes in a screen, letting it fall on a second screen placed behind the first. Upon illuminating a screen with two pinholes of light, he obtained diffraction patterns that consisted of alternating light and dark bands. He showed that in the region of overlap between the two beams, constructive (light) and destructive (dark) interference were taking place. This phenomenon could be explained only on the basis of the wave theory. Newton's discrete particle theory would not have exhibited the same patterns. A musician as well as a scientist, Young knew that sound waves of different frequencies could strengthen or cancel each other. His only explanation for the dark lines formed on the screen was that light waves from the different holes cancel each other out.

When he measured the waves, Young found that different colors have different frequencies, red having the lowest and violet the highest. Attempting to account for color vision, Young theorized that the retina harbors sensors that respond to different frequencies and hence to different colors of light. By experimentation he showed that all colors could be reproduced by combining red, green, and violet. Young therefore concluded that the retina needs only three types of sensors corresponding to these colors to produce color vision. In 1817 he proposed that the waves from which light is formed are transverse to the direction of propagation of the beam. This idea embodied the first satisfactory explanation of the polarization of light (because waves vibrating at right angles to their direction of travel could continue to vibrate even when restricted to one plane parallel to that direction).

Young's crucial contribution to the wave theory of light was his assumption that light is a transverse rather than a longitudinal wave. Transversality of the field means that the electric and magnetic field vectors lie in planes normal to the direction of propagation, as in light. Longitudinal disturbance means that the disturbance in the field is in the same direction as the direction of propagation, as in sound. Young based his idea on experiments with polarized light conducted by the French scientists Dominique-Francois-Jean Arago (1786–1853) and Augustin Jean Fresnel (1788–1827) in 1816. Fresnel and Arago found that two polarized rays at right angles to each other never interfere with each other, a discovery in conflict with the assumption that light is a longitudinal wave.

The classical particle controversy continued until the proponents of the corpuscular theory, encouraged by advances made by Pierre Simon de Laplace (1749–1827) and Jean-Baptiste Biot (1774–1862), suggested that the 1918 prize question sponsored by the Paris Academy should be to solve the problem of diffraction. The implication of their proposal was that the successful solution would be based on and confirm the "emitting corpuscle" theory. To their great surprise and dismay, the award went to Fresnel, whose treatment of diffraction was based on Huygen's wavelet envelope method and Young's interference principle. Fresnel's successful analysis of the diffraction phenomenon, based on his refinements of Huygens idea of secondary spherical wavelets, included proposal of an "obliquity factor," which concentrated the intensity of the wavelet toward the direction of propagation of the wave front. This factor was essential in correcting Huygens model because no backward-directed wave was observed. Fresnel's analytic technique was to split the individual spherical wave fronts into concentric rings of "half-period zones." The new wave front at a point in the forward direction of propagation was obtained by in-phase contributions of the concentric rings of the spherical wave.

Fresnel's brilliant application of these concepts led to concise descriptions of a number of phenomena. In 1818, on the basis of Fresnel's work, the French mathematician Siméon Denis Poisson (1781–1840) put forward an idea favoring Newton's corpuscular theory. He argued that if a shadow of a perfectly round object is made by a point source, then at the periphery of the object all the waves must be in phase and a bright point must thus be present at the shadow center. When Fresnel and Arago carried out the experiment, a bright point did appear at the center. It was another triumph for the wave theory.

Fresnel also initiated the analysis of dispersion phenomena in terms of a prism's molecular structural interaction with light waves. The separation of an electromagnetic wave into its various frequency components is called *dispersion*. A beam of white light, for example, can be separated into its monochromatic components because of the different velocities of rays at different wavelengths of the beam as it passes through a prism. The dispersion of a material, such as glass or water, at a given wavelength is defined as the rate of change of refractive index with wavelength. Fresnel's work was further elaborated by the French mathematician Augustin Louis Cauchy (1789–1857).

The most convincing evidence negating the corpuscular theory of light was put forward in 1850. According to the corpuscular theory, the corpuscules' attraction for dense media causes refraction (implying a faster speed of light in denser media), Jean Foucault (1819–1868) and Armand Fizeau (1819–1896) observed that light actually slows down when passing through water (as compared with air), a phenomenon consistent with only the wave refraction theory.

Joseph von Fraunhofer (1787–1826), a contemporary of Fresnel, made his contribution to the wave theory by examining a special case of diffraction patterns, when an object is situated at a long distance from the optical instrument. Fraunhofer showed that, for monochromatic light, the complex amplitude of the diffraction pattern is related to the Fourier transform of the entrance aperture. Even in the presence of aberration, a modified Fourier transform of the entrance aperture is related to the complex amplitude of the diffraction pattern. Fraunhofer's work formed the basis of Fourier optics and the development of spatial filtering techniques.

In 1876 Gustav Kirchhoff (1824–1887), using a method of integration superior to those previously employed, corrected two deficiencies in Fresnel's wave theory:

the obliquity factor and the phase of the spherical wavelets. Later the German-born American physicist Albert Abraham Michelson (1852–1931), devising a delicate experimental apparatus, refined the calculations of the speed of light to an accuracy of about 0.03%.

1.5 THE LINE SPECTRA OF ATOMS

Fraunhofer is also noted for establishing the phenomenon of absorption lines, known as *Fraunhofer lines*. After Newton initiated the study of spectroscopy, William Wollaston (1766–1828) continued it, discovering dark lines in the solar spectrum in 1802. From 1814 to 1817 Fraunhofer pursued the study of these absorption lines in the solar spectrum, finally obtaining their wavelengths. In 1861, more than forty years later, Robert Bunsen and Gustav Kirchhoff, experimenting with the solar spectrum, reconfirmed the existence of Fraunhofer lines. Their experiments indicated that the continuous spectrum emitted from the sun's core loses radiation in some wavelengths as it passes through the sun's atmosphere. Bunsen and Kirchhoff thus concluded that each atom has its own characteristic wavelength in emission and absorption.

These observations, which elucidated the process of line absorption and emission, seem to have had much more significance in the development of atomic theory than the observations made from Maxwell's equations. The intriguing question of how light and matter interact led to the theoretical development of the quantum theory in the twentieth century. The experiments of Fraunhofer, Bunsen, and Kirchhoff also gave impetus to further work in spectroscopy, through which the line spectrum of any chemical element can be detected and analyzed.

1.6 ELECTRICITY AND MAGNETISM

In the late 1700s, as experimental investigations in optics were beginning to bear fruit, major advances were being made in the field of electricity and magnetism. Fundamental discoveries and innovations in all three fields in the period 1775–1865 culminated in the reformulation of the classical equations of electromagnetism.

In 1785, after a series of impressive experiments, the French physicist Charles Augustin de Coulomb (1736–1806) formulated a law stating that the force between two electric charges is proportional to their product and inversely proportional to the square of the distance between them. He also showed that the force exerted on one charge is the (vector) summation of forces exerted by other charges present.

Next the Danish physicist and chemist Hans Christian Orsted (1777–1851) discovered in 1820 that an electric current flowing through a wire could deflect a magnetic needle. His experiment was crucial to electromagnetism because it showed that a flow of electricity produces a magnetic effect. Within a week of Orsted's discovery, André Marie Ampère (1775–1836), a French mathematician, developed a theory for this new phenomenon. Known as *Ampère's law*, it expressed mathematically the magnetic force between two electric currents. Ampère demonstrated that two adjacent wires, given freedom of movement, attract each other if their electric currents flow in the same direction. Conversely, the same wires repel each other if the currents flow in opposite directions.

Soon thereafter, in 1827, the German physicist Georg Ohm (1787–1854) found that a current flowing through a conductor is proportional to the electrical potentials at the two ends of the conductor and inversely proportional to the electrical resistance between the two ends of the wire. Although his discovery was given a cold reception by the scientific community of the time, it was eventually acknowledged as true and is now known as *Ohm's law*. Later, Karl Friedrich Gauss (1777–1855), a German mathematician, physicist, and astronomer, advanced the potential theory by formulating mathematical equations relating the electric field on a closed surface to the total electrical charge enclosed by that surface.

Michael Faraday (1791–1867), an English pioneer of science and one of the most imaginative experimentalists of all time, is known for two simple but enlightening discoveries. Aware of Orsted's experiment and convinced of the interrelationship between electricity and magnetism, in 1831 he demonstrated the phenomenon of electromagnetic induction and the generation of an electric current by a change (time variation) in the magnetic field. Faraday's experimental device consisted of two magnetic bars and a wire through which an electric current flowed. The current passing through the wire caused the magnetic bars to rotate. When the magnetic bars are fixed, the current passing through the wire causes the wire to rotate around the magnet. Faraday hypothesized that the current passing through a wire introduces a flux line or a magnetic field. He was thus the first to show that electricity and magnetism, although of different manifestation, originate in the same source. Faraday's experiments in electricity were the first to be applied to the practical mechanical and electrical innovations that would later revolutionize human life.

1.7 ELECTROMAGNETIC THEORY

Although the fields of electricity, magnetism, and optics seemed disjointed in the mid-eighteenth century, a century later it had become apparent that a vital connection existed among them. For more than two centuries, research in optics had persisted, despite a major controversy over the wave and corpuscular theories of light. The remarkable advances being made simultaneously in the fields of electricity and magnetism eventually led to their becoming merged in the field of electromagnetism. The mathematical formulas for electricity and magnetism developed during this period were proven by experimentation to hold true. These experiments suggested that electricity and magnetism, because of their common properties, might originate in the same source. The stage was thus set for the three fields to meet and combine forces, leading to the formulation of a unified and all-encompassing theory.

Scottish physicist James Clerk Maxwell (1831–1879) was recognized as a genius in his own time, and he succeeded in unifying these seemingly different disciplines. His brilliant theory gave mathematical form to several of the most fundamental physical and empirical relationships between electricity and magnetism. He expressed this theory in a single set of equations.

Inspired by Faraday's line of thinking, Maxwell took upon himself the task of investigating the equation of a basic relationship between the electric and the magnetic fields. He based his work on Coulomb's law, Gauss's theorems, Faraday's law, and Ampère's law. While formulating his equations, Maxwell realized that a significant term was missing from Ampère's law. Utilizing a symmetrical approach, he hypothesized that the generation of a magnetic field when an electric field has a time variation is analogous to the generation of an electric field when a time-varying magnetic field exists. Discovering the term missing from Ampère's law, he called it the *displacement current*. By integrating this term into his formulas, Maxwell produced his electromagnetic equations.

Maxwell unified all previous experimental results and their corresponding equations into a set of equations that took the form of a wave equation. He deducted the speed of propagation of this wave, c, from purely electrical and magnetic constants. Maxwell inferred that the light is also an electromagnetic wave, thus bringing together all three fields, a crucial turning point in the history of optics.

1.8 MAXWELL'S EQUATIONS

Maxwell's four equations describe all the phenomena of electricity and magnetism. They demonstrate that electricity and magnetism are not separable. *Wherever a time-varying electric field exists, there must be a magnetic field.* Maxwell's equations, relating the electric field intensity E, the magnetic field intensity H, the electric displacement D (flux density), the magnetic flux density B, the volume charge density ρ, and the current density J, are expressed as

$$\begin{aligned} \nabla \times E &= -\frac{\partial B}{\partial t} && \text{Faraday's law} \\ \nabla \times H &= J + \frac{\partial D}{\partial t} && \text{Ampère's law} \\ \nabla \cdot D &= \rho && \text{Gauss's law} \\ \nabla \cdot B &= 0 && \end{aligned} \tag{1.2}$$

where ∇ is the Laplacian vector operator.

These equations also indicate that a change in the electric field induces a change in the magnetic field, which in turn induces a change in the electric field, and so on. The two fields generate each other, and the resulting electromagnetic field progresses outward in all directions. On purely theoretical grounds, Maxwell demonstrated that an electromagnetic wave propagates as a transverse wave.

Maxwell's equations produced dramatic results. Light was proven to represent a disturbance that occurred in the form of an electromagnetic wave. These equations, rich in content and ranking in importance with Newton's gravitational equations, are of a different form. They are unique, for they describe a new type of physical phenomena. Their distinctive features appear in all equations of modern physics. Einstein described the formulation of Maxwell's equations as "the most important event in physics since Newton's time."

In spite of his rigorous approach, Maxwell's theory was not immediately accepted. In 1888 Heinrich Hertz, a German physicist, generated and detected radio waves, thus finally verifying the existence of electromagnetic waves. Hertz's experiments firmly established the fields of optics and electromagnetism on the basis of the wave theory set forth in Maxwell's electromagnetic equations. From that point on, the wave theory of electromagnetism and Maxwell's equations were widely accepted.

1.9 PHOTOGRAPHY AND MOTION PICTURES

Among the most significant contributions of optics during the nineteenth century was the invention of photography. Derived from the Greek words *photos* (light) and *graphos* (writing), photography began as a simple technique of recording visible images on light-sensitive materials. It has since developed into one of the most extensively used and most popular tools. From arts and sciences to entertainment and medicine, from still-life photography to motion pictures, the camera has penetrated areas heretofore unknown to humankind. Photography has reached into every facet of our lives, touching our culture and society with its ability to preserve the images of our times.

For centuries, scientists and artists tinkered with Leonardo da Vinci's camera obscura, a dark chamber or room with a hole in one wall through which images outside the room were projected onto the opposite wall, but not until the beginning of the nineteenth century did the first photographic processes take form. In 1826 Joseph Nicéphore Niepce (1765–1833), a French lithographer and inventor, produced the first photograph made with a camera by using the action of light on an asphalt solution.

The negative-positive photographic system in use today was developed by William Talbot (1800–1877), a British physicist. In 1835, by soaking light-sensitive paper alternately in solutions of common salt and silver nitrate, he produced a negative from which he could make a limitless number of positives simply by putting another sheet of light-sensitive paper against the negative and exposing it to light. Although many other entrepreneurs experimented with similar techniques, Talbot and

Louis Daguerre must be credited with developing the two processes that remain the foundations of modern photography.

By the late nineteenth century the camera was undergoing constant development. In 1913 Oskar Barnack (1879–1936), head of the Leitz experimental department in Germany, designed a camera that became the progenitor of the 35-mm photographic technique, including comprehensive camera and accessory systems for general and scientific photography.

Color photography was first introduced, though without much success, in the mid-nineteenth century. In 1859 James Clerk Maxwell, having established the three basic photographic colors, succeeded in reproducing other colors by combining red, green, and blue light in varying proportions. The foundation for color-print processes was laid ten years later when Ducos du Hauren and C. Clos of France developed the "subtractive color synthesis," or superimposition of dye images, as a means of producing pictures in color. The process adapted what nature provided, in that the human eye is selectively sensitive to color in terms of primary stimuli of red, green, and blue light in suitable proportions. Similarly, the color-film process records the proportion of red, green, and blue colors in their mixtures with three layers of color-sensitive dye plus a layer of black.

The early developments in popular photography coincided with the first venture into cinema, which also attracted wide attention. The invention of moving pictures was a significant turning point in the field of photography. So simple in nature and yet so rich in content, today's movies are created simply by projecting a series of photographs at a certain speed before a viewer. The required mechanical devices include reel-to-reel film, a projector, and a mechanism to move the film through the projector at different speeds to create slow motion or high speed and even stop action for viewing a single frame. The evolution of the first cinema machines was a gradual process, and proper applications of them were discovered slowly. Their potential began to be recognized during the 1920s, but without high-speed photography there could be no motion pictures.

Once photography and motion pictures became a high-technology category of their own during the 1960s, they were increasingly applied to scientific investigations. Today cameras are the leading optical instruments used in science and technology. One example of their value is seen in the study of biological procreation. Various high-performance photographic techniques have been especially useful in making the wonders of reproductive biology accessible to the public. Two of the most worthwhile educational uses of film today are in telescopic and microscopic viewing; Techniques to detect and project invisible or distant objects to the screen for clear viewing have dramatically transformed science and medicine.

Experiments with X rays, ultraviolet and infrared spectra, and the colors of visible light have made the use of film even more innovative. By employing the numerous spatial and temporal capabilities of film, we have learned a great deal about formerly invisible phenomena. Finely tuned camera equipment has made available different speeds and different magnifications for film, both of which have added another dimension to film making. The beauty and the precision of a blooming flower can be observed by viewing at high speed pictures that required hours of filming. The impact of a drop of water falling onto a surface can be observed by magnifying and slowing down a film of the event. Furthermore, by accelerating film taken by fiber-optic cameras, one can actually watch the miraculous process that leads from conception to a newborn infant. Through such techniques, heretofore imperceptible natural phenomena have become visible, exhibiting the drama of life in all its forms. Mysteries that confounded scientists for centuries have been solved through the magic of the camera.

1.10 OPTICAL PHYSICS IN THE TWENTIETH CENTURY

In the first decade of the twentieth century, three new theories in the field of optics restructured the foundations of classical physics and provided answers to three unsolved problems of the nineteenth century. The first theory, propounded by the German physicist Max Planck (1858–1947), concerned the laws of blackbody radiation. The next two theories, formulated by Albert Einstein (1879–1955), explored the photoelectric effect and the special theory of relativity.

Although these two scientists belonged to the old school of classical physics, their experimental findings could not be explained by its laws. By changing the foundations of classical physics, Einstein's theory modified the laws of classical mechanics and electromagnetism by way of the Lorentz transformation. The phenomena of blackbody radiation and the photoelectric effect could be explained only by assuming quantization of energy in the atom. This hypothesis was at odds with the assumption of continuous energy in the atom, which prevailed among scientists at that time.

Einstein's special theory of relativity forced a crucial shift in the development of relativistic physics. This theory is based on

1. The constancy of the velocity of light, as indicated by all previous measurements.
2. The axiom that Maxwell's equations must be invariant in form in all inertial frames.

In the following two postulates, Einstein presented his theory of relativity, which encompassed classical mechanics, electrodynamics, and all natural phenomena, resolving many of the attendant problems:

1. The laws of nature, including the laws of mechanics and the laws of electromagnetic phenomena, are the same in all inertial frames of reference, and they are independent of the translational motion of the system as a whole. More precisely, all inertial frames are completely equivalent for all phenomena.

2. The speed of light is independent of the motion of the emitting body. Light is propagated in empty space.

The first postulate is consistent with all experience in mechanics, where all relative motion between bodies is included, but it required modification under the laws of mechanics for very high speeds. The second postulate, untested when proposed and simple in nature, was decisively verified only in recent years. Some experiments have conclusively established the validity of Einstein's second postulate.

1.11 PLANCK'S THEORY OF RADIATION

Planck based his theory of radiation on the quantization of electron energy in an atom. According to his theory, the energy gain or loss of an electron passing from one energy level to another, higher, or lower than its initial level in an atom, must be discrete and is proportional to the frequency of the radiation emitted:

$$E_f - E_i = h\nu \tag{1.3}$$

where

- E_f is the electron energy in the final state,
- E_i is the electron energy in the initial state,
- ν is the frequency of radiation,
- h is the proportionality constant called *Planck's constant*. The magnitude of h was shown to be 6.625×10^{-34} W-s^2.

Planck, who was a learned thermodynamicist, initially applied the laws of thermodynamics to a blackbody and described its radiation behavior as thermal oscillation. He connected the thermodynamic energy of the blackbody to its electromagnetic energy through statistical mechanics. Planck concluded that the probability that a system would have a given energy level decreases exponentially with the magnitude of that energy level:

$$\frac{P_2}{P_1} = e^{-(E_2 - E_1)/kT} \tag{1.4}$$

where

- P_2 is the probability that the system has an energy level, E_2, corresponding to the second state,
- P_1 is the probability that the system has an energy level, E_1, corresponding to the initial state,

k is the Boltzmann's constant,
T is the absolute temperature.

Equation (1.4) indicates that the probability P_2 of a system having an energy level $h\nu$ reduces as the frequency ν increases. For an infinite frequency, the probability that the system has such an oscillation drops to zero.

Planck's novel theory predicted a substantial reduction in the blackbody spectrum $W(\nu, T)$ for high-frequency radiation. His famous radiation equations are

$$W(\nu, T) = \frac{2\pi h\nu^5}{c^3(e^{h\nu/kT} - 1)}$$
$$W(\lambda, T) = \frac{2\pi hc^2}{\lambda^5(e^{hc/\lambda kT} - 1)} \qquad (1.5)$$

where λ is the wavelength and is equal to c/ν. Although Planck's theory was compatible with experimental results, the scientific community, viewing it as a radical departure from the classical approach, preferred to ignore his results.

When Einstein postulated his theory of the photoelectric effect in 1905 (previously observed by Heinrich Hertz), he proposed that electromagnetic radiation itself be quantized. Einstein suggested that, when a quantum of electromagnetic radiation or a corpuscle (later to be called a *photon*) collides with one electron on the surface of some matter, a minimum of photon energy, called the *work function*, must be expended to remove the electron from the surface. Einstein postulated that each photon has an energy $h\nu$. If, on the one hand,

$$h\nu > \phi \qquad (1.6)$$

the electron escapes from the surface of the metal and the excess energy appears as the kinetic energy of the electron; that is,

$$h\nu = \phi + \frac{1}{2} m_e v^2 \qquad (1.7)$$

where

m_e is electron mass,
v is electron escape velocity.

If, on the other hand,

$$h\nu < \phi \qquad (1.8)$$

the electron cannot escape.

The Hertz-Einstein photoelectric effect could be explained only by the hypothesis of quantization of energy. However, because the wave theory was the dominant theory of light (owing to Maxwell's equations), the scientific community remained reluctant to accept the quantization of electromagnetic radiation.

In 1916 the American physicist Robert A. Millikan (1868–1953) undertook an investigation to disprove Einstein's experimental results of the photoelectric effect, but instead his work verified Einstein's theory. Thus the doors opened for the acceptance of the dualism concept of the wave-particle property of light. *The diffraction and interference of light were explained by the wave nature of light; and the photoelectric effect and the emission of blackbody radiation, by the quantum theory.*

In Planck's radiation law and Einstein's theory of the photoelectric effect, both based on the particle theory of light, the light-frequency term is explicitly expressed as the wave nature of light. This implied the inherent connection between the wave nature and the particle nature of light.

1.12 QUANTUM MECHANICS

In 1923, the French physicist Louis de Broglie (1875–1960), a scientist of insight and imagination, proposed a unified theory for photon and atomic particles. He observed a fundamental difference between particles and waves: the energy of a particle is concentrated in a small area of space confined to the particle, whereas waves distribute their energy in many directions and expand as a function of time. Furthermore, the trajectory of a particle follows the law of classical mechanics based on Newtonian laws, whereas waves undergo diffraction and interference that cannot be explained by the Newtonian mechanics of a trajectory. De Broglie, observing that these differences are obvious in the domain of classical physics (which primarily treats the macroscopic behavior of matter), believed that in the atomic domain particles may behave like waves and waves may behave like particles. By analyzing the comparison between waves and particles just described, de Broglie suggested his hypothesis: Associated with any material particle moving with momentum P, there is a wave of wavelength λ related to P according to the equation

$$\lambda = \frac{h}{P} \tag{1.9}$$

where h is Planck's constant.

Furthermore, according to de Broglie, associated with a particle such as an electron is a system of "matter waves" or "particle waves" that have slightly different velocities of propagation. At regular intervals along the path of propagation, the waves combine to form a crest that disappears at one point and reappears at another point in a later instant. The velocity of the crest, or "group velocity," is different from the velocities of the individual waves that combine to form it. De Broglie identified the group velocity as the velocity of the electron. The distance between successive crests in the de Broglie wavelength of the matter wave is

$$\lambda = \frac{h}{mv} \tag{1.10}$$

where

h is Planck's constant,
mv is the momentum of the particle.

In 1926 Erwin Schrödinger, an Austrian physicist deeply influenced by de Broglie's work, developed a time-wave theory for atomic particles, a stepping-stone in the field of quantum mechanics. Schrödinger's famous wave differential equation contains the energy of the particles as a parameter:

$$\nabla^2\psi(x, y, z) = \frac{-8\pi^2 m}{h^2}[E - U(x, y, z)]\psi(x, y, z) \tag{1.11}$$

where

∇ is the Laplacian operator,
E is the energy of the system,
U is the potential energy of the system.

In this equation, ψ is the wave function, which has a well-defined wavelength and phase velocity.

Schrödinger's wave theory, later substantially modified by other scientists, is the main support of the quantum theory. As the key formula in atomic and particle theory, it is one of the most important mathematical breakthroughs in physics in the twentieth century, ranking with Newton's eighteenth-century formulation of gravitation and Maxwell's nineteenth-century equations of electromagnetism.

1.13 OPTICAL DETECTORS

The development of quantum mechanics led to the birth of solid-state physics. As a result of advances in this new field and the ensuing exploration of light-sensitive elements, energy bands in various solid materials were recognized. The behavior of matter (particularly metals and semiconductors) under optical radiation was investigated. The energy bands in these materials and the corresponding wavelengths of optical radiation or emission were identified. When the energy bands in the intrinsic material could not produce a desired optical frequency, an additional element (impurity) was added to the host material. The impurity could be selected so that the corresponding energy bands created in the host material would match the desired optical frequency. The doped material ultimately obtained could be tuned to a variety of optical wavelengths. This flexibility in tuning a solid material to any desired optical frequency led to the production of numerous photodetectors.

The first reported optical detector developed (in 1935) was that of lead sulfide (PbS), used during World War II for missile guidance. Although many detectors were developed before 1950, their evolution has been mainly in the second half of the twentieth century, when research in photodetectors and their application, especially to optical instruments, sharply increased. During the 1950s individual detectors operating in the various optical spectra were fabricated and investigated.

Two significant developments in photodetectors were the indium antimonide (InSb) by Pieke in 1955, and the mercury cadmium telluride (HgCdTe) by Lawson in 1959, both of which enhanced the application of EO sensors. At the time electro-optical sensors were limited to the television type in the visible region and to reticle systems in the infrared region. The 1950s also witnessed the first application of a small array of detectors at the focal plane of an optical system.

During the 1960s emphasis was primarily on manufacturing linear array of detectors, which were incorporated into practical electro-optical systems and then tested. In the latter part of the decade, the technology of *integrated circuits* (ICs) was combined with the technology of the photodetector to create a hybrid integration of photodetectors and processing electronics. This combination, influenced by silicon, the common base material for both detector and processing, soon created a monolithic approach to so-called focal-plane technology. The period also saw the development of extrinsic doped-germanium detectors, such as Ge:Hg, and Ge:Cu, and doped-silicon detectors, like Si:As, operating in the 3–5 μm and 8–12 μm spectral bands.

During the 1970s the fabrication of both linear and two-dimensional mosaic arrays reached maturity. During the early part of the decade, almost all EO sensors were of the scanning type; therefore, linear arrays were popular. One invention, however, changed the fabric of the focal-plane detector. In 1970 a semiconductor called the *charge-coupled device* (CCD) was proposed by Boyle and Smith. CCD is a semiconductor material that can store the charge at one point and transfer and detect it at another point on command. Its performance includes a wide range of analog signal processing, as well as digital logic.

The already highly developed silicon technology was efficient enough for early fabrication of CCDs. The fabrication of compact and dense focal-plane arrays was a major step forward for detector technology in the latter part of the 1970s. In this period linear arrays of detectors on the order of a few thousand elements were assembled with *metal-insulator-semiconductor* (MIS) technology. Extrinsic Si detectors and Hg:Cd:Te detectors were widely used. As always, improvements in sensitivity and resolution in EO sensors were at a premium; innovative ideas were also flourishing in the design of EO sensors. One of these techniques was *time delay and integration* (TDI), which dramatically improved the design and the sensitivity of EO sensors. Another approach was to make a two-dimensional mosaic array similar to the human retina, thus eliminating the mechanical scan of a line array for coverage of the field of view. This new method sharply increased the sensitivity of sensors and led to the design of staring sensors.

Efforts in the early 1980s were focused on a modular focal plane that combines a detector with processing electronics, in either a hybrid or monolithic format, using intrinsic as well as extrinsic material. Today this technology has been advancing so rapidly that the fabrication of two-dimensional mosaic detectors consisting of millions of detectors and operating in the same way as the human retina is not beyond the realm of possibility.

Single, linear, and two-dimensional mosaic arrays for detectors, operating in the various spectral bands, have been developed, and research in this area of optics is moving forward rapidly. The increase in the sensitivity of these detectors and the reduction of their noise are accomplished by cooling them to cryogenic temperature. Liquid nitrogen (77 K), liquid hydrogen (20 K), and liquid helium (4 K) are used as coolants for detectors located at the focal plane of the optical device. In space, cooling is made possible by the proper baffling of optics and by taking advantage of the temperature of space (4 K). In some instances, active cooling elements have been incorporated.

1.14 ELECTRONICS AND SIGNAL PROCESSING

Concurrent with the development of semiconductor detectors, the transistor was invented in 1948 by John Bardeen and Walter Bratten, two American physicists working in the Bell Laboratories. In the same year the American physicist William Shockley's classic paper on junction diodes and transistors was published. The introduction of transistors immediately spurred development of electronic devices, thus changing the entire landscape of electronic circuitry. The convenient size, weight, and power requirements for a transistor, along with the flexibility of combining passive and active components, resulted in a powerful signal-processing circuitry.

Focal-plane detectors were followed by electronic and signal-processing circuits, which are either housed with the detector at the focal plane or installed as a separate module. When housed with detectors, these electronic elements can be fabricated monolithically with the detectors or connected to detectors as a hybrid component. As advances were made in fabricating sensitive detectors, the monolithic approaches were preferred.

Today the design and fabrication of electronic devices are being retransformed by *large-scale integration* (LSI), a process whereby tens of thousands of transistors and their interconnections are manufactured simultaneously. Through this technology, virtually all the logic elements of a digital computer are fitted onto a chip of silicon measuring only a quarter-inch on each side. Large-scale integrated circuits are fabricated with much smaller transistors, which can be packed more closely together than the older ones. The change in scale is measured crudely by counting the number of transistors fitted onto one chip. Small-scale integrated circuits have an average of 10 transistors per chip, medium-scale circuits have approximately 100, and LSIs have between 10,000 and 20,000. Today the number of transistors per chip is approaching 100,000. Since the late 1970s, VHSICs (very high-speed integrated circuits) have increased the availability of high-speed ICs. At the present the goal is to house large numbers of detectors (possibly thousands or even millions) with their corresponding processing of electronics in the focal plane of the optics.

1.15 FIBER OPTICS

Another development in the 1970s was fiber optics, in which a glass fiber or transparent plastic operates on the principle of total internal reflection. *Total internal reflection* refers to the reflection geometry at the boundary of two optically transparent but different media when the incident light in the first medium is reflected back to the same medium. Normally, the fibers are clustered together in a bundle that serves as a rod or pipe to transmit light. The light entering on one side of the rods or pipes can be detected from the other side. The light travels along the straight rods or even bends around a gentle corner, resulting in an internal reflection. With such reflective properties along the sides, there is very little loss of light. The device is particularly useful for viewing images produced at inaccessible locations; for example, physicians use it to examine internal organs of the body, often without surgery. The integration of fiber optics and mosaic detectors has made electro-optical systems highly efficient as well as flexible.

1.16 THE LASER

One of the major innovations in the field of optics was development of the laser in the 1960s. Although Einstein conceived the idea of the laser as early as 1917, almost half a century passed before it was realized for practical applications. In his 1917 paper, Einstein analyzed the process of light emission and absorption, using an approach that was a drastic departure from that of classical physics.

Einstein considered a simple atom having only two energy states, E_1 and E_2. If E_1 is the lower and E_2 the higher, the transition from one state to the other can be realized by emission or absorption of light quanta. If the probability per unit time of the atom falling spontaneously from the higher state to the lower state is A, and the probability per unit time of its passing from the lower state to the higher state under the influence of radiation is B, then, by requiring that the atom and the radiation be in statistical equilibrium, Einstein showed that the two rates are equal, A

= *B*. That is, the number of atoms per second going from the lower to the higher state is the same as the number going from the higher to the lower state. Through this concept he obtained the transition probabilities and the emissivity of the radiation, as well as a complete derivation of Planck's radiation law. His 1917 paper, called "*A* and *B* of Einstein" or "Einstein's Coefficients," is now recognized as a classic in physics.

Einstein's *B* coefficient became the foundation for the description of stimulated emission of radiation, which is the basis of laser radiation. In the stimulated emission process, an incident photon stimulates the atom to make a transition from a higher to a lower energy state. During the process photons of the same frequency are emitted. The stimulated emission from large numbers of atoms produces an intense light. Amplification of stimulated emission of radiation in an atomic system may be described by a combination of the Boltzmann law for energy distribution, the Einstein theory of radiation, and the selection rules in quantum mechanics.

The American physicist Charles Townes was the first to build an amplifier that used excited molecules as the active element (in this case, ammonia). The amplification was in microwave and hence the amplifier was called *MASER* (microwave amplification of stimulated emission of radiation). The theory of light amplification through stimulated emission of radiation (Laser) was put forward by Townes and A. Schawlow (also an American physicist) in 1954. The practical laser operation was achieved in 1960 by Theodore H. Maiman, who used a ruby rod as the lasing material. Since then a variety of materials have been tested as laser sources.

Laser beams revealed themselves as light sources of unprecedented intensity, collimated to a high degree. They can vaporize the hardest materials. The laser has initiated a revolution in optics and has permitted the observation of new phenomena, such as those of nonlinear optics and holography. Lasers promise to be the cutting and welding tool of the future, both in industry and medicine. They have already been used to perform microsurgery, such as welding the retina of an eye to its support. A wide variety of lasers (gas lasers, liquid lasers, and solid-state lasers) cover many regions of the electromagnetic spectrum. The intense and coherent quality of the radiation has led to increasing application of lasers in fields such as radio astronomy, microwave spectroscopy, photography, biophysics, and communications.

1.17 HOLOGRAPHY

Holography was initially proposed in 1948 by Dennis Gabor, a Hungarian-born British physicist, who discovered that if the diffraction pattern of an object is recorded on a photograph plate the image of the original object can be reconstructed from the plate. Although Gabor's discovery had merit, it did not reach fruition until the laser was developed. Emmett N. Leith and Juris Upatniek reflected laser beams from various objects, combined them with the original beam, and recorded the results on a photographic plate. The image was then reconstructed by illuminating the plate with a laser source. Thus far holography has had limited application in industry. Its main advantage is its three-dimensional photographic capability, which remains to be exploited.

1.18 ELECTRO-OPTICAL SENSORS

The period from 1900 to 1980 was marked by the union and evolution of optics and electronics, generating the complex field of electro-optics. The discovery of the photoelectric effect, which was instrumental in opening up the atomic world, was probably the most important factor in uniting optics with electronics. By inserting a light-sensitive detector at the focal plane of a converging optical instrument, such as a television camera, an electrical signal whose amplitude was proportional to the incident radiation was produced. Such signals were then transmitted to a remote location for conversion into optical radiation and display of the original scene to human observers. Thus the first visual sensing of distant objects in real time became a reality. It is noteworthy that both quantum mechanics and television, the first practical EO sensor, were developed in the same period (1900–1980).

As noted earlier, the design and analysis of EO sensors are very complex. The subject not only demands simultaneous application of optics and electronics, but also detailed knowledge of many other fields, including solid state physics, heat transfer, signal processing, data transmission, display, atmospheric phenomenology, and human engineering.

In recent years the application of electro-optical sensors in satellites and spacecraft for detection and transmission of weather data, identification of Earth's resources, and determination of composition and atmosphere of other planets has had striking results. Remote sensing devices, consisting of an electro-optical sensor operating in the infrared or ultraviolet spectrum, are now operational. These sensors include, for example, television-type cameras, ultraviolet and infrared sensors mounted on a satellite or aircraft for continuous mapping of designated areas on Earth or other planets.

The infrared region has been especially useful in night-vision devices and space probes that sense thermal images. Of particular value have been thermal imaging sensors operating in the middle wavelength of infrared (i.e., in the range of 3–5 μm bands) and in the long wavelength of infrared (i.e., in the range of 8–12 μm bands). There is considerable absorption of infrared radiation by the atmosphere owing to water vapor (H_2O), carbon dioxide (CO_2), ozone (O_3), and methane (CH_4), among other atmospheric compounds. Most infrared sensors therefore operate in the atmospheric windows, free from atmospheric absorption. The infrared spectrum is used in space probes to measure temperature and determine the composition of an alien planet by the spectral signatures of its component materials.

During the past 30 years the design of optical instruments has been dramatically improved by the use of computers. The variation of optical image quality as a function of the field of view and spectral bands can be evaluated through computer programs. In the past, a telescope's field of view was small, as looking at celestial bodies did not require a large field of view. At present, however, the extensive application of sensors in aircraft and space probes requires wide-angle fields of view and high image quality optics, so large refractive elements (lenses) must be developed. Large optical elements made of glass, silicon, and germanium have been tested and used as lenses. To maintain a high image quality in optics despite temperature variations, corrective elements have been included in the design. The positions of these elements are monitored by microprocessors. Larger fields of view have been achieved when a scanning mechanism is incorporated in the design. To accomplish this task, either the entire telescope must be rotated around an axis or a mirror installed inside the optical device to provide the desired scanning motion.

EO sensors were initially limited to line scanners, which dominated their design in the 1950s. During the 1960s and early 1970s the most sensitive EO sensors consisted of a few hundred detectors in a linear array assembled in a low-temperature Dewar housing. The signal output of each detector was then passed through the Dewar wall to a preamplifier housed outside it. The signal then passed from the preamplifier to a signal-conditioning and filtering multiplexer to prepare it for either visual display or for automatic data processing.

The need for higher sensitivity forced a different approach to EO sensor problems in the middle to late 1970s. Not only the detector but also the preamplifier and many other signal-conditioning circuitries were housed in the focal plane. The implementation was eased by the invention of CCDs in the early 1970s. In particular, *forward-looking infrared* (FLIR) imaging sensors, developed in the same period, incorporated a few hundred detectors as a linear array assembled by using thin-film technology.

1.19 ASTRONOMY AND THE DEVELOPMENT OF LARGE OPTICAL SENSORS

Discoveries and inventions in the field of astronomy are among the outstanding contributions of optics to the advancement of science and hence to the enrichment of human life. Curiously enough the universe has encouraged innovation in magnifying celestial bodies to advance knowledge of rays of light and the functions they perform. Now, in the twentieth century, the concept of light rays has expanded to embrace a wide variety of rays, ranging from radio waves to cosmic rays.

The Sumerian view of the universe was a simple one, holding that Earth, flat and fixed in space, was the center of the universe. The ultimate defeat of the geocentric view awaited evidence gathered by Galileo in the early 1600s. His observations confirmed the Copernican picture and laid the foundations for modern astronomy.

Advances in astronomy in the twentieth century outstrip those of previous eras. Large telescopes, photographic plates that can be exposed to radiation from the cosmos for long periods of time; ultraviolet, infrared, and X rays; microwaves and radio waves—all have contributed to the opening of a new world. Among the discoveries following these advances are external galaxies, nebulas, novas, and supernovas.

In the 1930s radio telescopes were invented, capable of forming images from radio waves emitted by celestial objects. More recently, optical telescopes have been modified to operate in ultraviolet and infrared regions of the optical spectrum.

Present astronomical observations utilize telescopes whose mirrors measure 100 to 200 in diameter. Most of them are installed atop high mountains to reduce disturbances caused by Earth's atmosphere. With the aid of such telescopes, the gas clouds in the Milky Way were discovered actually to be other galaxies, probably tens of billions in number and certainly at enormous distance from Earth. This discovery expanded the boundaries of the universe by a quantum leap, just as the presumed limits of the solar system had been expanded not long before.

In the 1920s the American astronomer Edwin Hubble (1889–1958), by using the new Mt. Wilson 100-in telescope, ascertained that the universe is receding in all directions. When a star is moving away from the observer, the absorption or emission lines of elements are displaced toward the longer wavelength of the spectrum. The displacement is called the *red shift*. By studying the spectra of galaxies, Hubble discovered the Doppler effect in their light waves, which is compatible with the red shift. He also found that the galaxies' speed of recession increases with distance, a fact known as Hubble's law.

From these observations emerged the Big Bang theory of cosmic evolution, which holds that at one time all matter in the universe was clustered together. Then, between 15 and 18 billion years ago, an explosion of nearly infinite intensity broke up the clustered mass, thus starting a continuous expansion of the universe. The cooling-down effect, after about a billion years, condensed the matter into separate galaxies and stars. Although scientists are in general agreement about the broad outlines of the Big Bang theory, a still puzzling problem is the total amount of mass in the cosmos, which will determine whether the universe will expand forever or the expansion will be slowed and finally reversed by gravitational forces. If there is enough mass, contraction and compression will prevail, and the universe might end in a "big crunch," that would obliterate all matter 50 to 100 billion years hence.

1.20 ADVANCES IN TELESCOPES

At present the world's largest telescope, with a 6-meter aperture, is located on Mt. Pastukhov in the Caucasus Mountains in the former Soviet Union. Before its installation, the largest was the 200-in telescope on Mt. Palomar in California, which is still in operation. Among other large telescopes are the 100-in one on Mt. Wilson in California and the 60-in one on Haleakala on the island of Maui in Hawaii.

Presently scheduled improvements in astronomical observation are a 100-in space telescope and a 400-in ground telescope. The former, named for the great astronomer Hubble, was launched in 1990. This telescope, however, encountered difficulties primarily from a spherical aberration of its mirror due to manufacturing deficiencies.

The 400-in ground telescope, named after its donor W. M. Keck, is mounted on Mauna Kea, Hawaii, at a height of 14,000 ft above sea level. It is to be operational in the 1990s. Figure 1.2 gives the general layout of the telescope and sup-

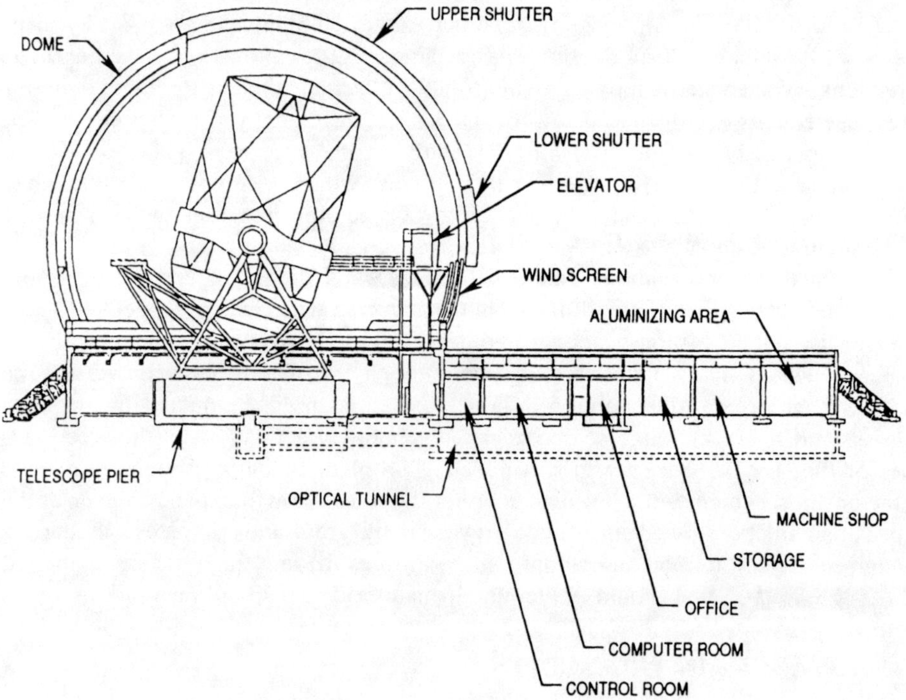

Figure 1.2 General layout of W. M. Keck telescope and supporting facilities.

porting facilities. Figure 1.3 is the primarly mirror support system. The Keck telescope is the first ground telescope whose design is radically different from other ground telescopes primarily because of its size.

As it is not possible to make an accurate 10-m mirror from a single slab of glass, the telescope will have a mosaic of tightly packet hexagonal mirrors, each 3-in thick, as shown in Figure 1.4. The 35 mirrors in the finished assembly will be aligned with utmost accuracy, within a fraction of wavelength, through a computer controller. The number of computer adjustments may reach a maximum of 300 per second.

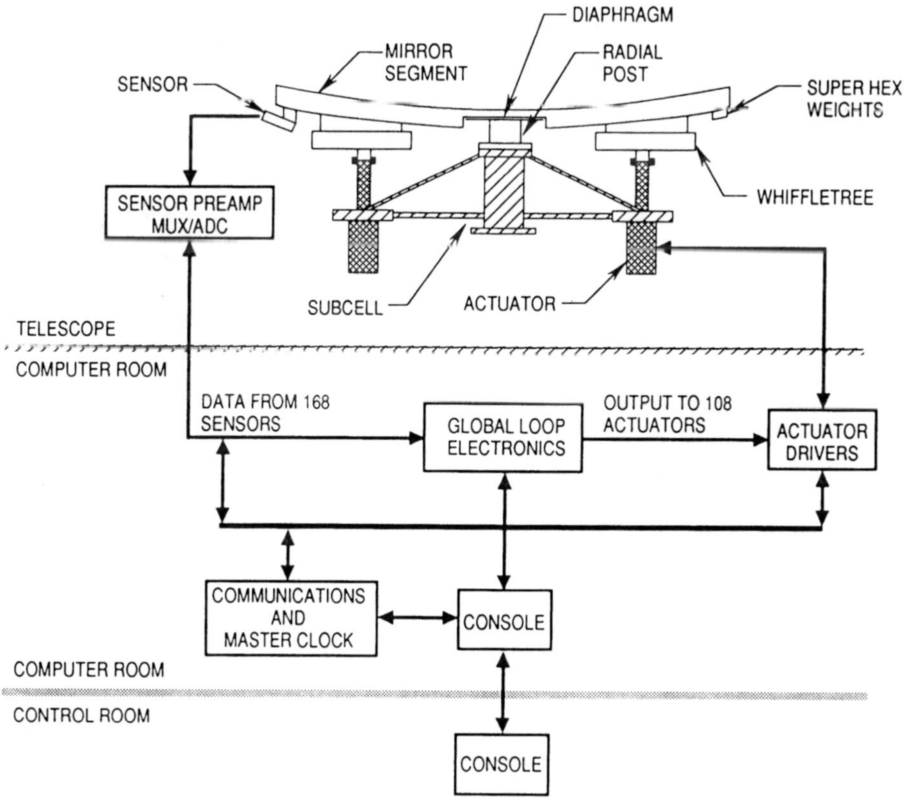

Figure 1.3 Primary mirror support system of W. M. Keck telescope.

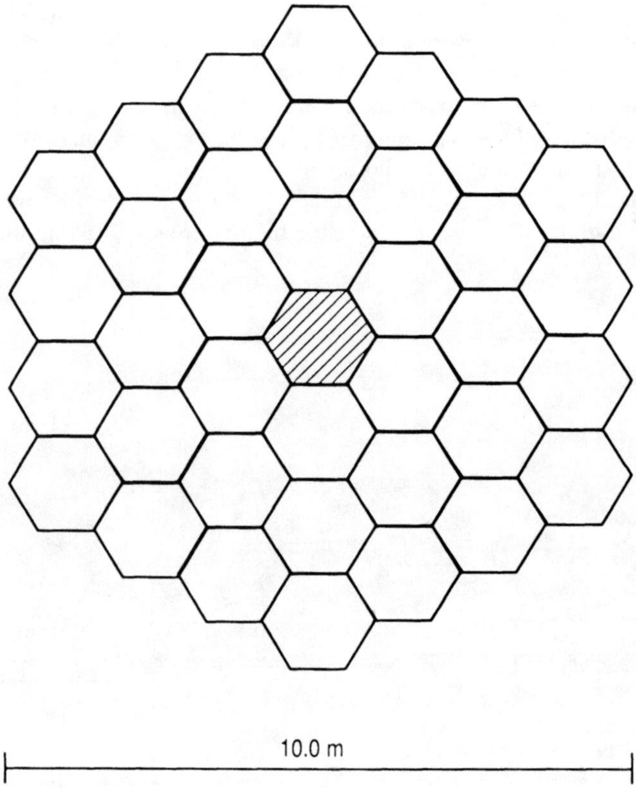

Figure 1.4 The 36-segment mirror geometry.

1.21 CONTEMPORARY OPTICAL SENSORS

Multiple detectors in a variety of configurations—including linear, staggered, two-dimensional mosaic arrays in ultraviolet, infrared, and visible spectra—are being incorporated into the focal plane of the optical sensors. Digital processing, encompassing large data-rate processing in very small chips, has been the major factor in the development of EO sensors. In recent years electro-optical detection and imaging sensors have ensured an excellent capability in remote sensing procedures for detection providing good picture quality. Implementing these sensors in aircraft and space probes has produced real-time pictures from other continents and other planets. Thus the new field of EO systems and satellite communication has made us aware of what exists beyond the range of human vision and in other spectral regions.

Electro-optical imaging systems are produced in a variety of spectral bands from visible to near and far infrared. For example, the Thematic Mapper used in the Landsat satellite contained seven bands as given in Table 1.1. Bands 1, 2, and 3 of Table 1.1 correspond to blue, green, and red portions of visible spectrum; bands 4 and higher correspond to the infrared spectrum. Each band produces specific results when imaging the Earth. Imaging spectrometers are constructed with a finer resolution of bandwidths. In these systems the spectral band of 0.4 to 2.5 μrad is divided into 200 subbands, to extract additional information from imaged areas.

Table 1.1
Thematic Mapper Spectral Band Characteristics

Range (μm)	*Characteristic*
0.45–0.52	Coastal water mapping
0.52–0.60	Healthy vegetation
0.63–0.69	Chlorophyll absorption
0.76–0.90	Water delineation, vegetation vigor
1.55–1.75	Snow or cloud differentiation
10.4–12.5	Plant-heat stress measurement
2.08–2.35	Hydrothermal mapping

Figure 1.5 shows several electro-optical scanning configurations for imaging ground terrain. In Figure 1.5(a), fixed detector elements are scanned across the surface of the Earth perpendicular to the flight path by a mirror, these detectors convert the reflected solar photons from each pixel in the scene into an electronic signal. The detector elements are placed behind filters that pass broad portions of the spectrum.

The primary limitation of this approach is the short dwell time of the detector in each *instantaneous field of view* (IFOV). To achieve adequate *signal-to-noise* (S/N) ratio without sacrificing spatial resolution, such a sensor must operate in broad spectral bands (100 nm or greater) or must scan a slow rate to reduce the area covered per unit time.

One approach to increasing the residence time of a detector in each IFOV is to use line arrays of detector elements (Figure 1.5(b)). In this configuration there is a dedicated detector element for each cross-track pixel, which increases the dwell, (or integration) time to the interval required to move one IFOV along the track. Such an experimental, two-spectral channel instrument, called the *modular optoelectronic multispectral scanner* (MOMS), has been carried aboard the space shuttle. A French satellite sensor (the Systeme Probatoire d'Observation de la Terre, or SPOT), launched in 1986, also uses line array detectors. The SPOT system provides stereoscopic image capability in three spectral bands for wavelengths of less than 1.0 μm.

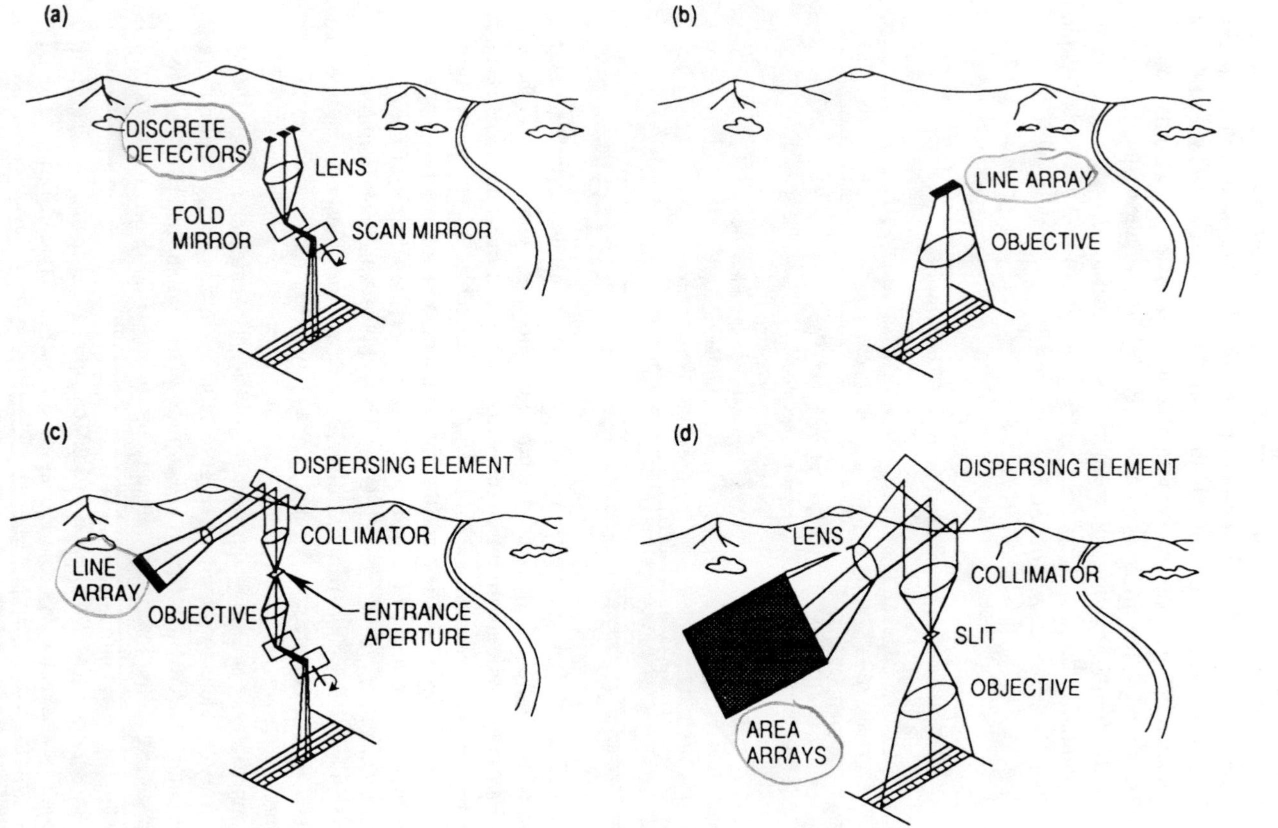

Figure 1.5 Four approaches to sensors for multispectral imaging: (a) multispectral imaging with discrete detectors; (b) multispectral imaging with line arrays; (c) imaging spectrometry with line arrays; and (d) imaging spectrometry with area arrays. (From [15].)

Some limitations and trade-offs are associated with the use of multiple line arrays, each having its own spectral band pass filter. If all the arrays are placed in the focal plane of the telescope, then the same ground locations are not imaged simultaneously in each spectral band. If beam splitters are used to facilitate simultaneous data acquisition, the signal is reduced by 50% or more for each additional spectral band acquired in a given spectral region. Furthermore, instrument complexity increases substantially if more than 6 to 10 spectral bands are desired.

Two other approaches to imaging spectrometry are shown in Figures 1.5(c) and 1.5(d). In the line array approach (Figure 1.5(c)), the light from a pixel is passed into a spectrometer, where it is dispersed and focused onto a line array. Thus each pixel is simultaneously sensed in as many spectral bands as there are detector elements in the line array. For high spatial resolution imaging (ground IFOVs of 10–30 m), this approach is suited only to an airborne sensor that flies slowly enough that the readout time of the detector array is a small fraction of the integration time. Because of the high spacecraft velocities, imaging spectrometers designed for Earth orbit require the use of two-dimensional area arrays of detectors at the focal plane of the spectrometer (Figure 1.5(d)), thereby obviating the need for the optical scanning mechanism. In this configuration there is a dedicated column of spectral detector elements for each cross-track pixel in the scene [15].

Figure 1.6 shows a typical optical ray trace of the optical beam from an outside object to the detector. The image is transmitted from the primary to the secondary

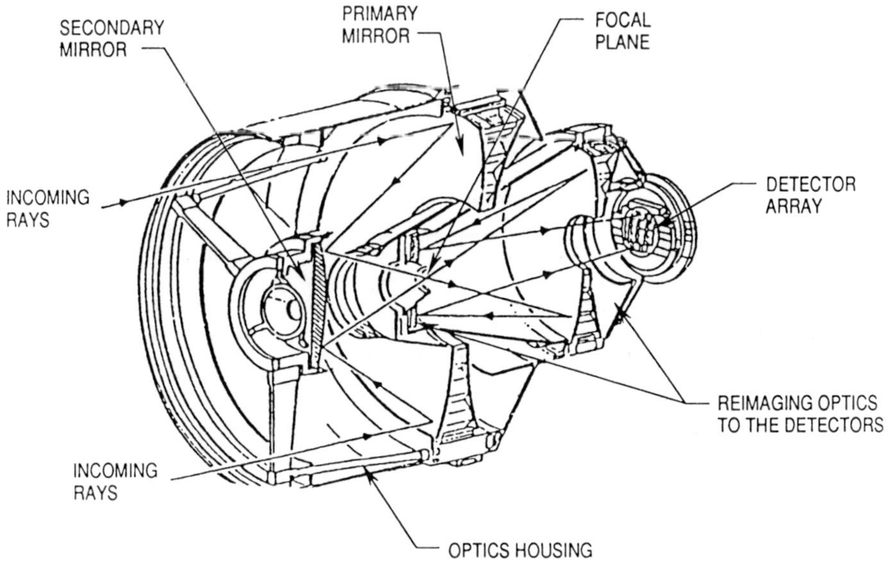

Figure 1.6 Ray trace of optical beam from an outside object to the detector.

mirror and then to the focal plane. Reimaging optics from the focal plane to the detector array is used to reduce indirect (stray) illumination paths. Figure 1.7 is a schematic diagram of the Galileo telescope, incorporating 800 × 800 CCDs, which was constructed for the Jupiter mission.

The aperture size of EO sensors ranges from about an inch in a typical photographic camera to a maximum of 10 m, the size in the Keck telescope. For a large aperture and small field of view most optical designs are reflective, whereas for a large field of view, about 10° or more, most designs are refractive. When the desired spectral bands are wide, the reflective design is superior to the refractive because it is free of chromatic aberration and lens absorption.

The typical detector size ranges from about 0.5 to 10 mil (1/1000 of an inch) and is usually in the neighborhood of 2 mil. The number of detectors ranges from 1 to as high as 10^6. To improve its sensitivity, a detector is kept cool by housing it in a Dewar container, where a cold finger is cooled by a thermoelectric cooler or cold gases. Liquid nitrogen, helium, and hydrogen are used for cooling. Cooling temperatures from 20 K to 77 K are typical for infrared detectors. In the visible region, however, cooling is not required.

As an example of electro-optical imagery obtained from space, Figure 1.8 shows a typical image of Earth's surface taken by the French SPOT satellite from an altitude of 828 km. The SPOT system operating in green, red, and near infrared bands con-

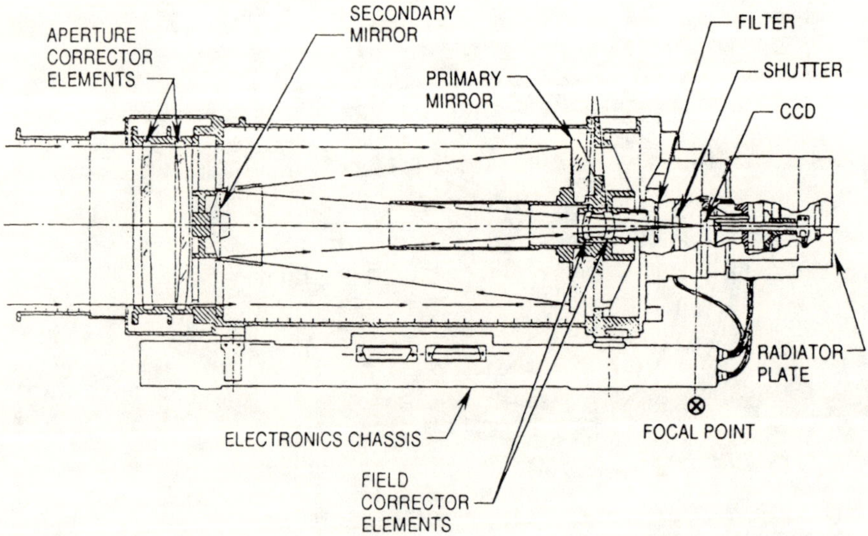

Figure 1.7 Galileo telescope optics configuration.

Figure 1.8 Image of Earth produced by SPOT satellite, bodies of water and rivers are shown in black; streets and urban areas are clearly visible.

sists of two high-resolution camera telescopes equipped with charge-coupled devices that provide high-resolution images. Each telescope measures 2.5 m in length and each CCD detector has a ground resolution of 10 m. Thus each row of 6,000 CCDs can produce a 60 km picture of the landscape below.

REFERENCES

[1] Hecht, E., and A. Zajac, *Optics*, Reading, MA: Addison Wesley, 1979.

[2] Born, M., and E. Wolfe, *Principles of Optics*, New York: Pergamon Press, 1959.

[3] Boorstin, D.J., *The Discoverers*, New York: Random House, 1983.

[4] Seyrafi, K., *Electro-Optical System Analysis*, Los Angeles: Electro-Optical Research Company, 1986.

[5] Seyrafi, K., *Light: The Mystery of the Universe*, Los Angeles: Electro-Optical Research Company, 1986.
[6] Jackson, J.D., *Classical Electrodynamics*, New York: John Wiley and Sons, 1975.
[7] Eisenberg, R., and R. Resnick, *Quantum Physics of Atoms, Molecules, Solids, Nuclei, and Particles*, New York: W.W. Norton Company, 1968.
[8] Resnick, R., *Introduction to Special Theory of Relativity*, New York: John Wiley and Sons, 1968.
[9] Van der Waerden, B.L. (Ed.), *Sources of Quantum Mechanics*, New York: Dover Publications, 1967.
[10] Weber, R.L., *Pioneers in Physics*, London: Institute of Physics, 1980.
[11] Townes, C.H., and A.L. Schanlon, *Microwave Spectroscopy*, New York: McGraw-Hill, 1955.
[12] Frazier, K., *Solar Systems*, Alexandria, VA: Time-Life Books, 1985.
[13] Eicher, D.J., "Keck's First Light," *Astronomy*, April 1991.
[14] Hovanessian, S.A., *Introduction to Sensor Systems*, Norwood, MA: Artech House, 1988.
[15] *High Resolution Imaging Spectrometer (HIRIS)*, National Aeronautics and Space Administration, 1987.

Chapter 2
Optical Radiation

2.1 INTRODUCTION

Optical radiation covers the *ultraviolet* (UV), visible, and *infrared* (IR) portions of the electromagnetic spectrum; that is, 10^{-2} μm to 10^{3} μm. Figure 2.1 shows the electromagnetic spectrum and the corresponding wavelengths. The ultraviolet portion ranges from about 0.1 to 0.38 μm. The visible portion is from approximately 0.38 to 0.76 μm in wavelength. The infrared portion is divided into the near infrared or *short-wavelength infrared* (SWIR) region (from 0.77 to 3 μm), the *middle-wavelength infrared* (MWIR) region (from 3 to 8 μm), the *long-wavelength infrared* (LWIR) region (from 8 to 14 μm), and the far and extreme infrared regions (from 14 to 1000 μm).

The radiant energy received by an optical system can be generally broken down into two components: (1) radiant energy due to the self-emission of objects and (2) radiant energy reflected or transmitted by objects due to external radiation sources. The objects being observed by the optical system can be categorized as either the target or the background. Figure 2.2 shows the typical components of radiation sources of an electro-optical system. The target is the object to be detected by the sensor; it must be distinguished from the background sources. The target can be considered as the optical signal. The background consists of all radiation sources received by the EO sensor, excluding the object of interest. They can include the terrain, clouds, ocean, atmosphere, and so on. The background can be considered as a form of noise, as will be discussed in Chapter 4.

The radiation emitted and reflected from the target and the background traverse through the atmospheric medium, where absorption and scattering take place. The transmitted radiation energy, then, is received by the EO sensor. Chapter 3 describes the transmission of the optical radiation through the atmosphere.

In the next section, the quantities of radiation measurements with their units and definitions are given.

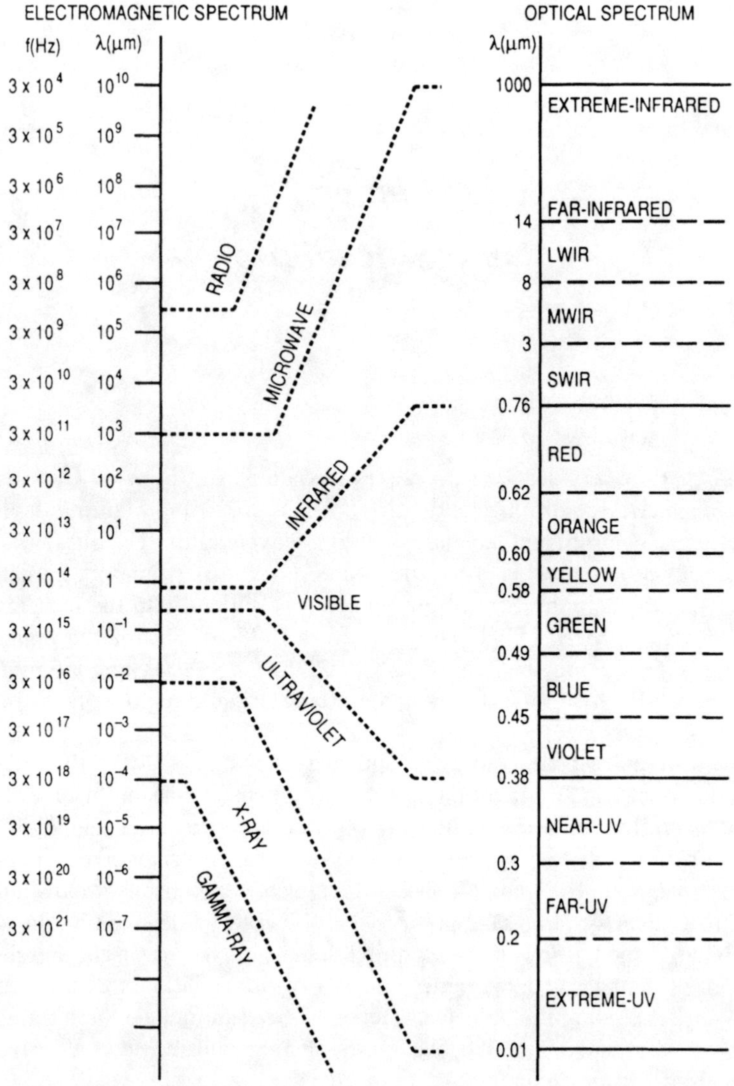

Figure 2.1 The electromagnetic spectrum.

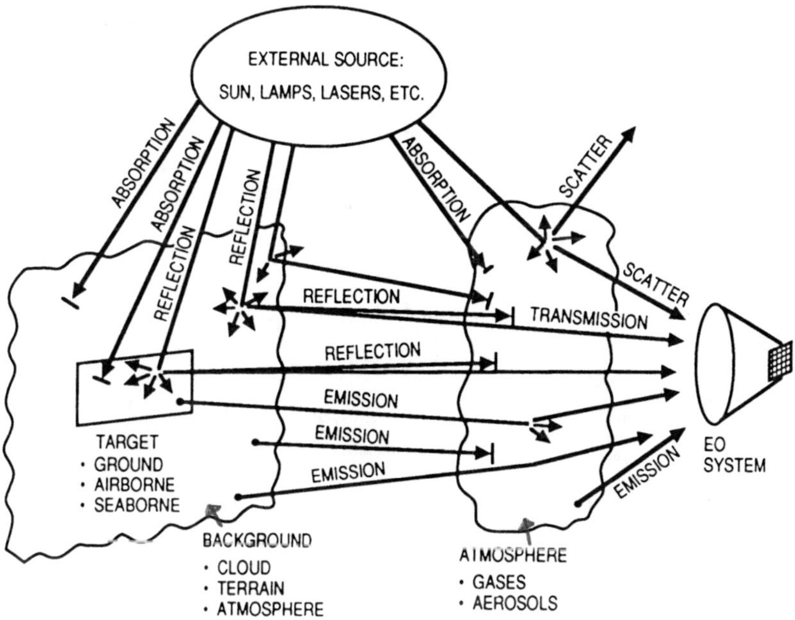

Figure 2.2 Typical radiative components of an EO system.

2.2 QUANTITIES OF RADIATION AND UNITS

The basic quantity in the measure of optical radiation is the radiant flux (power), designated by the symbol Φ or Φ_e. It is defined as the radiant power traversing a surface, given by

$$\Phi_e = \frac{\partial Q_e}{\partial t} \text{ Watts (W)} \tag{2.1}$$

where Q_e is the total radiant energy contained in a radiation field in Joules. The subscript e denotes energy-derived units.

The radiant exitance M_e, sometimes referred to as the *radiant emittance*, is defined as the radiant flux per unit area leaving a surface, expressed by

$$M_e = \frac{\partial \Phi_e}{\partial A} \text{ W/cm}^2 \tag{2.2}$$

where A is the area of the surface of a source of radiation as shown in Figure 2.3.

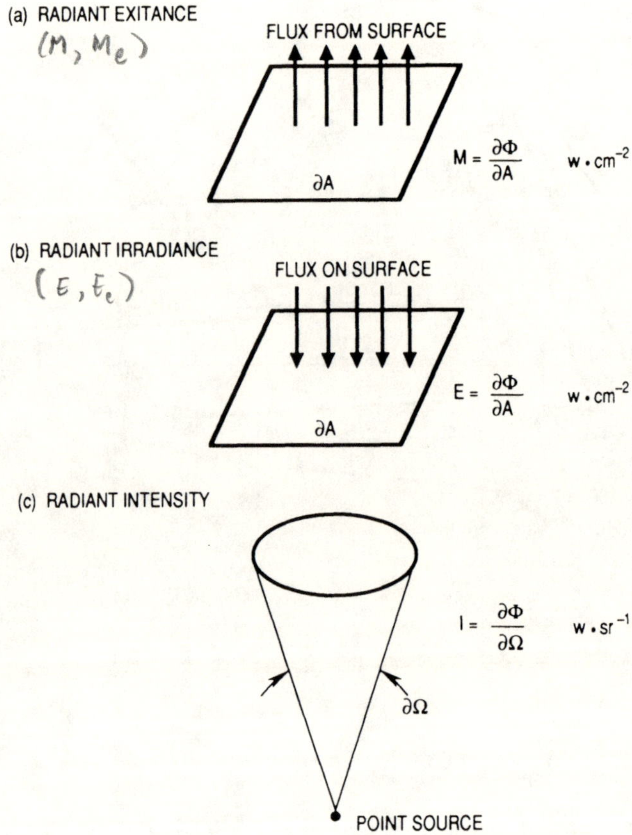

Figure 2.3 Quantities of radiation.

The radiant flux received by a unit area of a surface from another surface is called *irradiance*, E_e, also defined as

$$E_e = \frac{\partial \Phi_e}{\partial A} \text{ W/cm}^2 \tag{2.3}$$

The radiant intensity, I_e, and the radiance, L_e, are two additional radiation quantities defined in terms of radiant flux per unit solid angle. The solid angle, Ω, is defined as the three-dimensional angular spread at the vertex of a cone given by

$$\Omega = \frac{A}{r^2} \text{ steradian (sr)} \tag{2.4}$$

(d) RADIANCE

$$L = \frac{\partial^2 \Phi}{\partial A \cos\theta\, \partial\Omega} \text{ W} \cdot \text{cm}^{-2} \cdot \text{sr}^{-1}$$

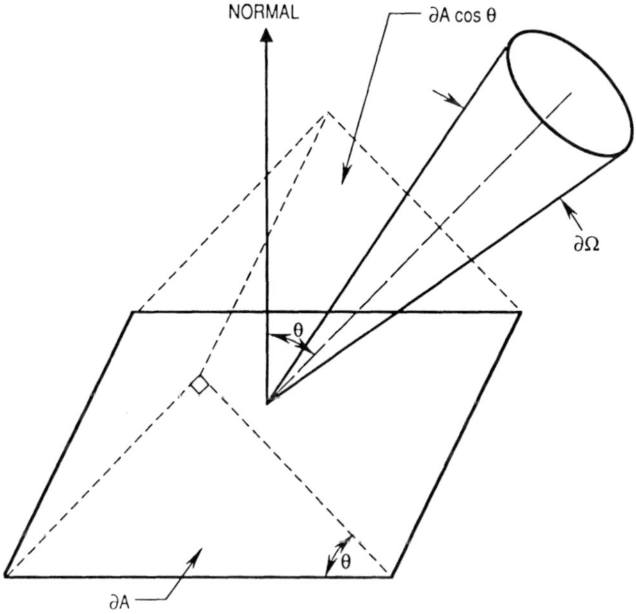

Figure 2.3 Quantities of radiation (continued).

where A is the spherical surface area intercepted by the cone, and r is the radius of the sphere (see Figure 2.4(a)).

Equation (2.4) is valid for spherical surfaces only; but as $r^2 \gg A$ or as Ω gets smaller, the area of a nonspherical surface is closely approximated by a spherical surface area. In the most practical instances, $r^2 \gg A$ and eq. (2.4) can be used. Because the surface area of a sphere is $4\pi r^2$, the solid angle subtended by an entire sphere, from eq. (2.4) is 4π sr. For a hemisphere, the subtended solid angle is 2π sr.

If the half vertex angle $\theta/2$, shown in Figure 2.4(b), of a right circular cone is known, then the solid angle is given by the expression

$$\Omega = \pi \sin^2\left(\frac{\theta}{2}\right) \text{ sr} \tag{2.5}$$

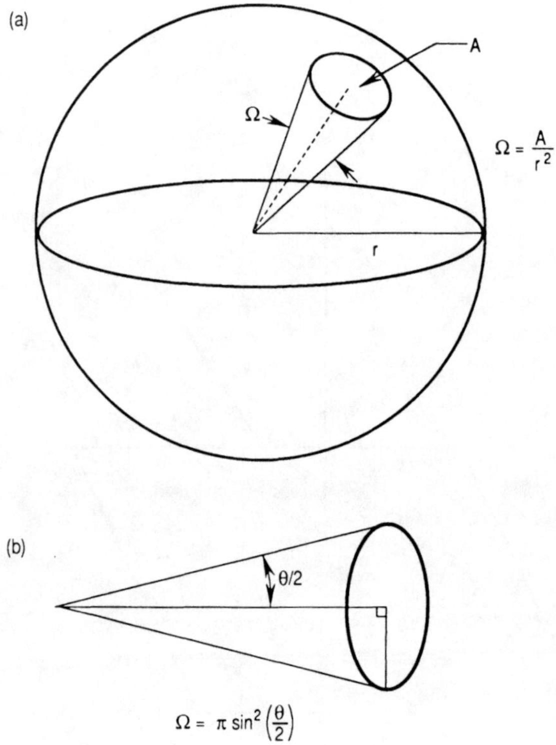

Figure 2.4 Definition of a solid angle.

The radiant intensity I_e, defined as the radiant flux per unit solid angle leaving the entire source of radiation, is given by

$$I_e = \frac{\partial \Phi_e}{\partial \Omega} \text{ W/sr} \tag{2.6}$$

The radiant intensity is particularly useful in the determination of the amount of radiation produced by a point source object.

The radiance, L_e, is defined as the radiant flux per unit solid angle per unit projected area leaving a surface. It is expressed as

$$L_e = \frac{\partial I_e}{\partial A \cos\theta} = \frac{\partial^2 \Phi_e}{\partial A \cos\theta \partial \Omega} \text{ W-cm}^{-2}\text{-sr}^{-1} \tag{2.7}$$

where

∂A	is the surface area of radiation,
$\partial A \cos\theta$	is the projected area as shown in Figure 2.3(d),
θ	is the angle between the normal of the surface area and the line of sight.

The quantities of radiation measurements can also be expressed in terms of the photon flux Φ_p, defined as the number of photons emitted, transferred, or received per unit time. It is given by

$$\Phi_p = \frac{\partial Q_p}{\partial t} \text{ photons/s} \tag{2.8}$$

where Q_p is the photon energy given by

$$Q_p = \frac{Q_e}{h\nu} = \frac{Q_e \lambda}{hc} \text{ photons (ph)} \tag{2.9}$$

where λ is the wavelength, h is the Planck's constant, ν is the frequency, and c is the speed of light. The subscript p here denotes photon-derived units. Other photon quantities—that is, photon exitance, photon irradiance, photon intensity, and photon radiance—can be defined in terms of the photon flux, Φ_p. Table 2.1 lists the radiometric quantities given by eqs. (2.1) to (2.7) and also the photon quantities based on the photon flux.

Table 2.1
Radiometric and Photon Quantities

Quantity	*Symbol*	*Units*	*Definition*
Radiant energy	Q, Q_e	J	joules
Radiant flux (power)	Φ, Φ_e	W	$\partial Q_e/\partial t$
Radiant exitance	M, M_e	W-cm^{-2}	$\partial \Phi_e/\partial A$
Irradiance	E, E_e	W-cm^{-2}	$\partial \Phi_e/\partial A$
Radiant intensity	I, I_e	W-sr^{-1}	$\partial \Phi_e/\partial \Omega$
Radiance	L, L_e	W-sr^{-1}-cm^{-2}	$\partial^2 \Phi_e/\partial A \cos\theta \partial \Omega$
Photon energy	Q_p	photons (ph)	Basic unit
Photon flux	Φ_p	ph-s^{-1}	$\partial Q_p/\partial t$
Photon exitance	M_p	ph-s^{-1}-cm^{-2}	$\partial \Phi_p/\partial A$
Photon irradiance	E_p	ph-s^{-1}-cm^{-2}	$\partial \Phi_p/\partial A$
Photon intensity	I_p	ph-s^{-1}-sr^{-2}	$\partial \Phi_p/\partial \Omega$
Photon radiance	L_p	ph-s^{-1}-sr^{-1}cm^{-2}	$\partial^2 \Phi_p/\partial A \cos\theta \partial \Omega$

2.3 BLACKBODY RADIATION

When light is incident upon an object, the following terms can be used to describe its behavior:

1. Absorptance (α)—the ratio of the absorbed radiant power to the incident radiant power;
2. Reflectance (ρ)—the ratio of the reflected radiant power to the incident radiant power;
3. Transmittance (τ)—the ratio of the transmitted radiant power to the incident radiant power.

Any combination of the processes—absorption, reflection, and transmission—can take place on the object. Fractions of the total incident radiant energy are absorbed, reflected, or transmitted, so that

$$\alpha + \rho + \tau = 1 \tag{2.10}$$

In addition to these processes, which are due to external radiation sources, the object itself can emit radiation, as illustrated in Figure 2.2.

A perfect mirror has the properties $\rho = 1$ and $\alpha = \tau = 0$. A perfect window has the properties $\tau = 1$ and $\alpha = \rho = 0$. If an object has the properties $\alpha = 1$ and $\rho = \tau = 0$, then that object is called a *blackbody*.

A blackbody is defined as an ideal body or surface that absorbs all radiant energy incident upon it at any wavelength and at any angle of incidence, so that none of the radiant energy is reflected or transmitted. Therefore, the term *blackbody* refers to a perfect absorber of radiation. Because all radiant energy is absorbed by a blackbody and none is reflected or transmitted, it will appear to be "black." An approximation of a true blackbody is a hollow cavity with blackened inner surface and a small aperture where light can enter. Any radiation entering the aperture will undergo many low-order reflections inside the cavity until nearly all radiation is absorbed.

Kirchhoff's law of electromagnetic radiation states that a good absorber is also a good emitter of radiation by the equation:

$$\alpha = \varepsilon \tag{2.11}$$

where

- α is the absorptance,
- ε is the emissivity defined as the ratio of the radiant power emitted per unit area from a surface to the radiance emitted per unit area from a blackbody.

Because all radiation incident on the surface of a blackbody is absorbed, $\alpha = 1$ and its emissivity is also equal to unity, that is, $\alpha = \varepsilon = 1$ for a blackbody. Therefore, the term blackbody not only refers to a perfect absorber of radiation but also to a perfect emitter of radiation; one that radiates the maximum number of photons possible per unit time per unit area in a specific spectral interval and temperature. Two important emission properties of a blackbody are

1. It is a perfect emitter of radiation at all angles and all wavelengths,
2. The total emitted energy of a blackbody is a function only of its temperature.

The concept of a blackbody serves as a standard of comparison. It is the ideal radiator with which we can compare sources of real objects. Emissivity, ε, is the term used for comparison of real objects to a blackbody. A blackbody, by definition, has an emissivity of unity. A radiation source with an emissivity less than unity is known as a *graybody*.

Every object in nature is a source of radiation. The peak energy of this radiation will depend on the object's temperature and will be concentrated in the infrared, visible, or ultraviolet region of the optical spectrum. As the temperature of an object increases, the total radiated energy of the object will increase with a corresponding shift in the peak of the radiation wavelength toward a shorter wavelength.

The total radiated energy from a unit area of a surface in a unit of time over a 2π steradian solid angle is called the *radiant exitance* of that surface. For blackbody sources, the radiant exitance $M_e(T)$, is proportional to the fourth power of the absolute temperature of the source; that is,

$$M_{e,\mathrm{BB}}(T) = \sigma_e T^4 \tag{2.12}$$

where

σ_e is the Stefan-Boltzmann constant and has a value of 5.6687×10^{-12} W-cm^{-2} K^{-4},
T is the absolute temperature of the source in degrees Kelvin.

In these units, $M_e(T)$ is therefore in W-cm^{-2}. Equation (2.12) for a graybody is written as

$$M_{e,\mathrm{GB}}(T) = \varepsilon \sigma_e T^4 \tag{2.13}$$

and ε is the emissivity.

Similarly, the photon exitance, $M_p(T)$, for a blackbody is proportional to the third power of the absolute temperature of the source, that is,

$$M_{p,\mathrm{BB}}(T) = \sigma_p T^3 \tag{2.14}$$

where

$$\sigma_p = 1.52 \times 10^{11}\ \text{ph} \cdot \text{s}^{-1} \cdot \text{cm}^{-2} \cdot \text{K}^{-3}$$

2.4 PLANCK'S RADIATION FORMULA

The preceding discussion of the radiant power of a blackbody, using the Stefan-Boltzmann law, was based on thermodynamic principles. The Stefan-Boltzmann law gives the total radiant power of a blackbody as a function of temperature. However, this law does not describe the nature of light radiation or its spectral radiance distribution. The wavelength dependency of radiation and the total energy available in a specified band interval from an object are prerequisites in evaluating sensor irradiance and the effect of background on detector performance in most optical systems.

The spectral radiant exitance of a surface is defined for unit wavelength. If the wavelength is given in micrometers (μm) the spectral radiant exitance is in W-cm^{-2}-μm^{-1}. Although in the dimensions of spectral radiant exitance, the surface area and the spectral band both have dimensions of length (or length^2), they are almost always expressed separately to indicate the object radiant power from a unit surface area within a unit spectral band.

Spectral dependence and the nature of light radiation can be analyzed with the quantum theory of radiation. Planck's radiation law applies this theory and can be used to calculate the spectral radiance of blackbody sources. Planck's equation satisfies thermodynamic requirements and gives the spectral radiant exitance $M_{e,\mathrm{BB}}(\nu, T)$ of a blackbody in terms of temperature, T, and radiation frequency, ν:

$$M_{e,\mathrm{BB}}(\nu, T) = \frac{2\pi h\nu^5}{c^3(e^{h\nu/kT} - 1)} \tag{2.15}$$

where

- h is Planck's constant,
- c is the speed of light,
- k is Boltzmann's constant.

Equation (2.15) can also be written in terms of wavelength, λ:

$$M_{e,\mathrm{BB}}(\lambda, T) = \frac{2\pi hc^2}{\lambda^5(e^{hc/\lambda kT} - 1)} \tag{2.16}$$

Substituting $\lambda = c/\nu$ in eq. (2.15) will yield $M_{e,\mathrm{BB}}(\lambda, T)$.

Equations (2.15) and (2.16) express the spectral radiant exitance of a source propagated in a vacuum. For any other media, c should be replaced by the light velocity in the media, c/n_r, where n_r is the refractive index of that media.

Figure 2.5 shows the variation of $M_{e,\mathrm{BB}}(\lambda, T)$ as a function of λ and T. Equation (2.16) indicates that for very large or very small values of wavelength, the function

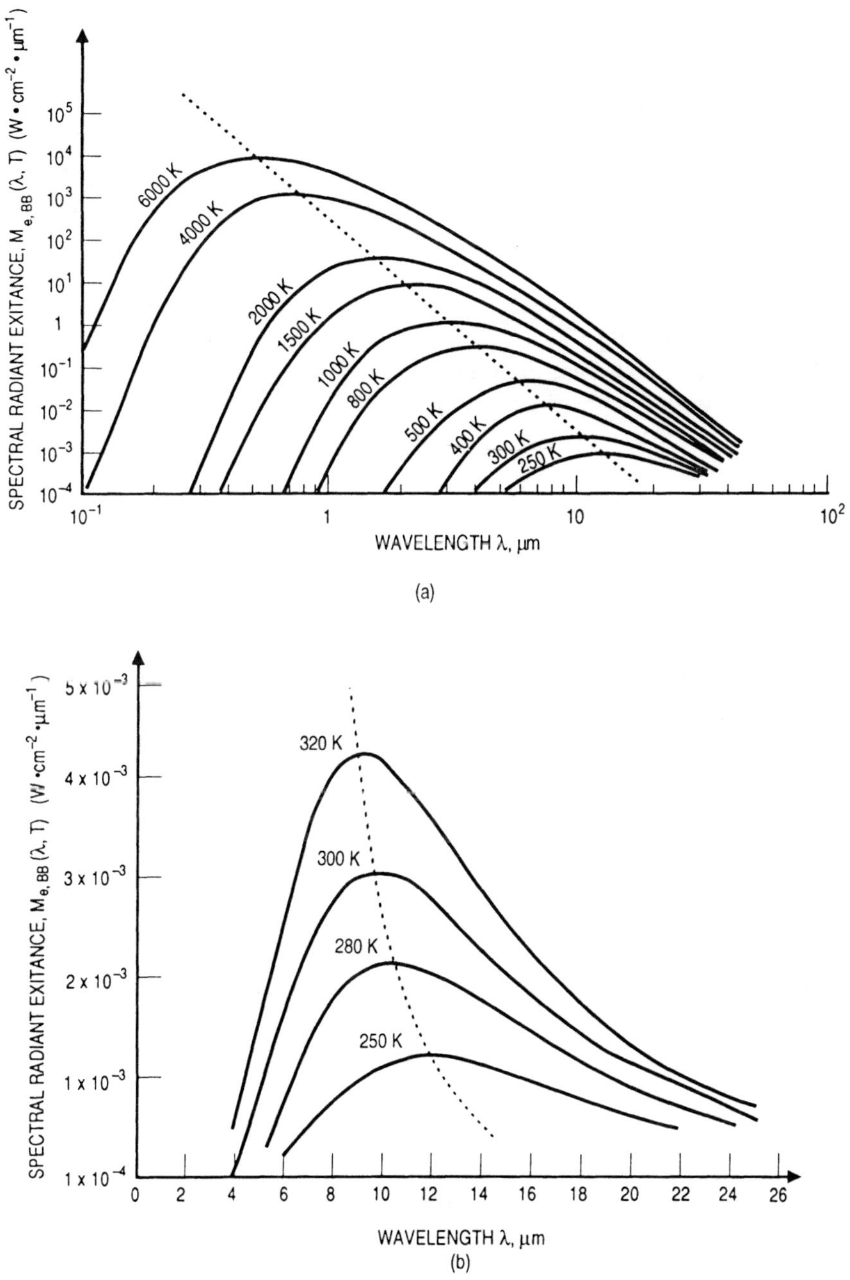

Figure 2.5 (a) Absolute radiant exitance and (b) absolute spectral radiant exitance of blackbodies at various temperatures.

$M_{e,BB}(\lambda, T)$ goes to zero. The integral of eq. (2.16) over all possible λ, that is, $\int_0^\infty M_{e,BB}(\lambda, T)d\lambda$, gives the Stefan-Boltzmann equation, eq. (2.12), for blackbody radiation.

The wavelength for the peak of the spectral radiation can be calculated by using Planck's formula. Taking the derivative of eq. (2.16) with respect to λ and evaluating its zero:

$$\lambda_{max}T = 2897 \ \mu\text{m-K} \tag{2.17}$$

This equation indicates that the wavelength of the maximum radiation shifts toward a shorter wavelength as the temperature of a source increases. This is called *Wein's displacement law.* Figure 2.6 gives the peak wavelength, λ_{max}, of a radiation source as a function of source temperature.

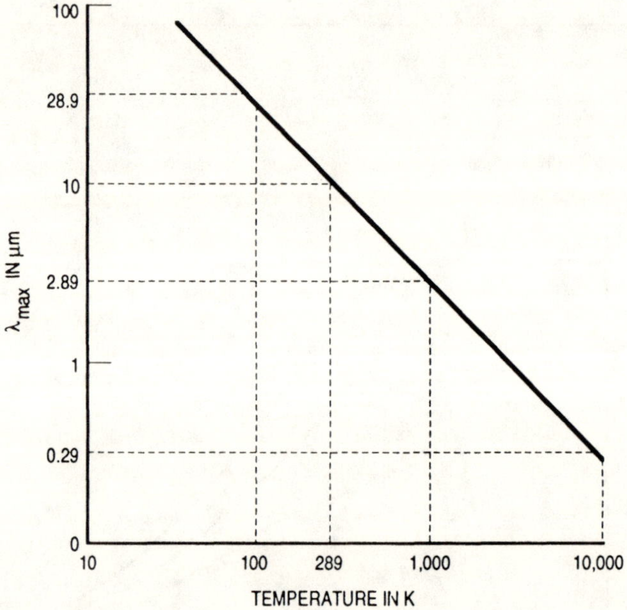

Figure 2.6 Peak radiation wavelength of a blackbody source as a function of its temperature.

For a graybody, the spectral radiant exitance, $M_{e,GB}(\nu, T)$, in terms of temperature, T, and radiation frequency, ν is

$$M_{e,GB}(\nu, T) = \varepsilon(\nu)\frac{2\pi h\nu^5}{c^3(e^{h\nu/kT} - 1)} \tag{2.18}$$

where ε is emissivity. In terms of wavelength, λ, eq. (2.18) can be written

$$M_{e,\mathrm{GB}}(\lambda, T) = \varepsilon(\lambda) \cdot \frac{2\pi hc^2}{\lambda^5(e^{hc/\lambda kT} - 1)} \tag{2.19}$$

The photon exitance of a blackbody or a graybody can be calculated by dividing eqs. (2.18) and (2.19) by the energy of one photon, that is, $h\nu$. For a blackbody, the photon exitance is

$$M_{p,\mathrm{BB}}(\nu, T) = \frac{2\pi\nu^4}{c^3(e^{h\nu/kT} - 1)} \tag{2.20}$$

or

$$M_{p,\mathrm{BB}}(\lambda, T) = \frac{2\pi c}{\lambda^4(e^{hc/\lambda kT} - 1)} \tag{2.21}$$

Figure 2.7 shows the variation of $M_{p,\mathrm{BB}}(\lambda, T)$ as a function of λ for several temperatures. The integral of eq. (2.21) over all possible wavelengths, that is, $\int_0^\infty M_{p,\mathrm{BB}}(\lambda, T)d\lambda$, results in eq. (2.14) for blackbody radiation.

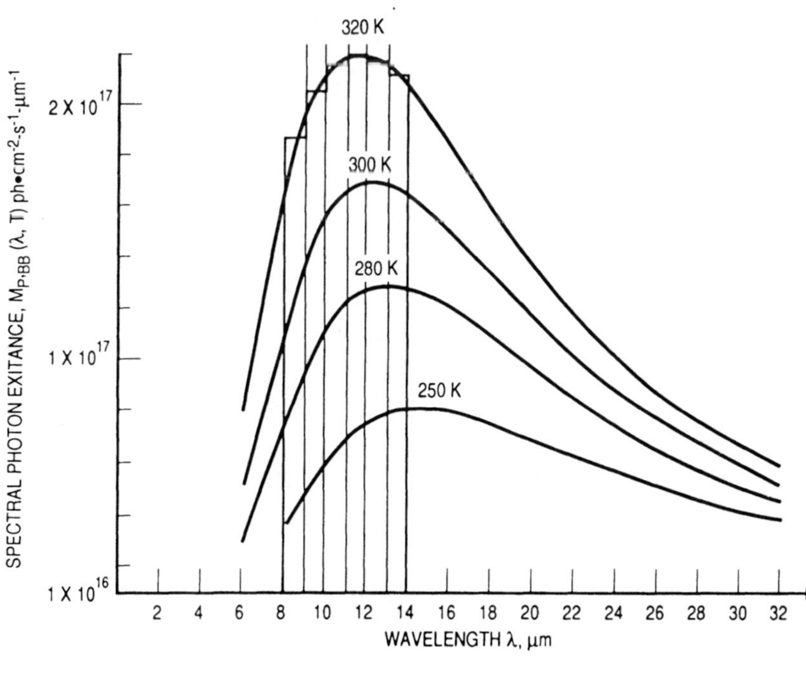

Figure 2.7 Spectral photon exitance of blackbodies at various temperatures.

Similarly, for a graybody, the photon exitance is given by

$$M_{p,\mathrm{GB}}(\nu, T) = \varepsilon(\nu) \cdot \frac{2\pi\nu^4}{c^3(e^{h\nu/kT} - 1)} \tag{2.22}$$

or

$$M_{p,\mathrm{GB}}(\lambda, T) = \varepsilon(\lambda) \cdot \frac{2\pi c}{\lambda^4(e^{hc/\lambda kT} - 1)} \tag{2.23}$$

Example

Evaluate the spectral radiant exitance and the spectral photon exitance of a graybody with emissivity $\varepsilon = 0.3$ at 10 μm. The temperature of the body is 300 K.

First we evaluate the spectral radiant exitance of a blackbody at 10 μm. According to blackbody formula, for $T = 300$ K the spectral radiant exitance at $\lambda =$ 10 μm is

$$M_{e,\mathrm{BB}}(10\ \mu\mathrm{m}, 300\ \mathrm{K}) = \frac{2\pi hc^2}{\lambda^5(e^{hc/\lambda kT} - 1)}$$

$$= \frac{2\pi(6.626 \times 10^{-34})(3 \times 10^{10})^2}{\lambda^5\left[e^{\dfrac{(6.626 \times 10^{-34})(3 \times 10^{10})}{\lambda(1.38 \times 10^{-23})T}} - 1\right]}$$

$$= \frac{3.74 \times 10^{-12}}{\lambda_{\mathrm{cm}}^5[e^{1.4388/\lambda_{\mathrm{cm}}T} - 1]}\ \mathrm{W\text{-}cm^{-2}\text{-}cm^{-1}}$$

$$= 31.16\ \mathrm{W\text{-}cm^{-2}\text{-}cm^{-1}}$$

$$= 3.116 \times 10^{-3}\ \mathrm{W\text{-}cm^{-2}\text{-}\mu m^{-1}}$$

For graybody with emissivity of $\varepsilon = 0.3$, the radiant exitance at $\lambda = 10$ μm is

$$0.3 \times 3.13 = 0.94 \qquad \mathrm{mW\text{-}cm^{-2}\text{-}\mu m^{-1}}$$

The photon exitance of a 300 K blackbody at 10 μm is

$$M_{p,\mathrm{BB}}(10\ \mu\mathrm{m}, 300\ \mathrm{K}) = \frac{2\pi c}{\lambda^4(e^{hc/\lambda kT} - 1)}$$

$$= \frac{1.884 \times 10^{11}}{\lambda_{\text{cm}}^4 \, (e^{1.4388/\lambda_{\text{cm}}T} - 1)} \text{ ph-s}^{-1}\text{-cm}^{-2}\text{-cm}^{-1}$$

$$= 1.57 \times 10^{21} \text{ ph-s}^{-1}\text{-cm}^{-2}\text{-cm}^{-1}$$

$$= 1.57 \times 10^{17} \text{ ph-s}^{-1}\text{-cm}^{-2}\text{-}\mu\text{m}^{-1}$$

For one μm bandwidth at $\lambda = 10$ μm the photon exitance is

$$M_{p,\text{BB}} \, (10 \, \mu\text{m}, 300 \text{ K})|_{\Delta\lambda = 1\mu\text{m}} = 1.57 \times 10^{17} \text{ ph-s}^{-1}\text{-cm}^{-2}\text{-}\mu\text{m}^{-1}$$

For graybody with emissivity of $\varepsilon = 0.3$ the photon exitance is

$$M_{p,\text{GB}} \, (10 \, \mu\text{m}, 300 \text{ K})|_{\Delta\lambda = 1\mu\text{m}} = 4.71 \times 10^{16} \text{ ph-s}^{-1}\text{-cm}^{-2}\text{-}\mu\text{m}^{-1}$$

Example

Calculate the photon exitance of a 320 K graybody in the spectral interval 8–14 μm. Assume the emissivity in this band to be constant at 0.4.

The photon exitance over the spectral interval 8–14 μm is given by the integral

$$M_{p,\text{GB}} \, (T) = \int_8^{14} \varepsilon(\lambda) \, M_{p,\text{BB}} \, (\lambda, T) \, d\lambda$$

Because the emissivity is assumed to be constant at 0.4 over the 8–14 μm band,

$$M_{p,\text{GB}} \, (T) = 0.4 \int_8^{14} \frac{2\pi c}{\lambda^4 (e^{hc/\lambda kT} - 1)} \, d\lambda$$

The integral is the area under the 320 K curve in Figure 2.7 in the spectral interval from 8–14 μm. This is approximately

$$1.26 \times 10^{18} \text{ ph-s}^{-1}\text{-cm}^{-2}$$

Therefore,

$$M_{p,\text{GB}} \, (320 \text{ K})]_{8-14\mu\text{m}} = 5.04 \times 10^{17} \text{ ph-s}^{-1}\text{-cm}^{-2}$$

The detailed mechanism of the emission and absorption of radiation energy is beyond the scope of this book. However, the essential equations in the evaluation of the radiance of an object or the irradiance at the entrance aperture of an optical sensor are discussed in the following paragraphs.

2.5 RADIATION FROM A SURFACE

The radiance of a surface in a given angular direction Z_θ is defined as the amount of energy radiated per unit time, per unit solid angle, per unit of projected area of the surface normal to the direction Z_θ. This radiated energy may be due to surface self-emission, reflections from the surface, or transmission through the surface from an external illuminating source.

If we take an element dS_1 of a surface S_1, as shown in Figure 2.8, the differential power radiated $d\Phi$ in Watts, from dS_1, in the solid angle $d\Omega$ in the direction Z_θ, will be

$$d\Phi = L_\theta \, d\Omega \, dS_1 \cos\theta \tag{2.24}$$

where L_θ is the surface radiance in the Z_θ direction in W-cm^{-2}-sr^{-1}.

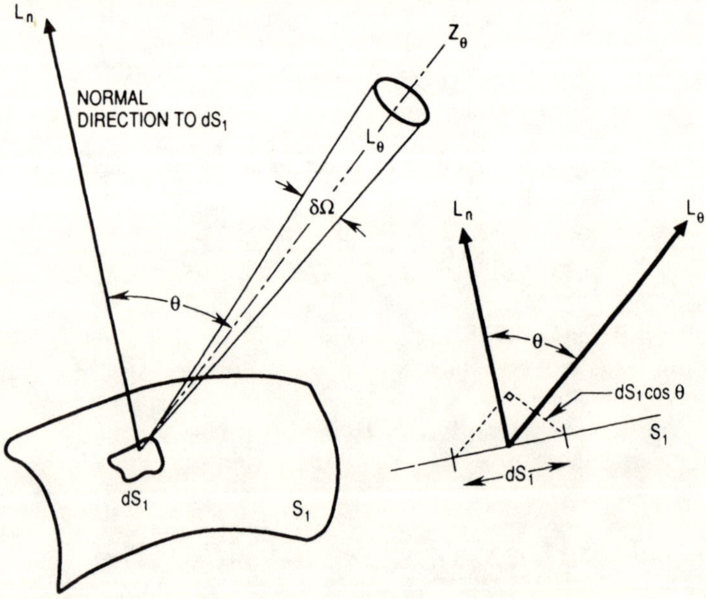

Figure 2.8 Geometry of a radiant power of an element dS_1 into a solid angle $d\Omega$.

The total power, Φ, radiated from the surface S_1 in the solid angle $d\Omega$ in the direction Z_θ is

$$\Phi = d\Omega \int_{S_1} L_\theta \cos \theta dS_1 \tag{2.25}$$

In general, L_θ is a function of θ. The surface is called diffusive or Lambertian if the radiance L_θ is the same in all directions. The radiant power per unit solid angle per unit area (not projected area), $\partial^2\Phi/\partial\Omega\partial S$, of a Lambertian surface in a given direction, Z_θ, decreases with the cosine of the angle between the Z_θ direction and the plane normal to the surface:

$$\left(\frac{\partial^2\Phi}{\partial\Omega\partial S}\right)_\theta = L_n \cos\theta \tag{2.26}$$

where

- L_n is the surface radiance or surface brightness in the direction normal to the surface,
- θ is the angle between the plane normal to the surface and the direction of the radiation.

Figure 2.9 shows the radiant power per unit solid angle per unit area of a diffusive surface as a function of the angle θ. Equation (2.24) for a blackbody surface can be written as

$$d\Phi = L_n \, d\Omega \, dS_1 \cos\theta \tag{2.27}$$

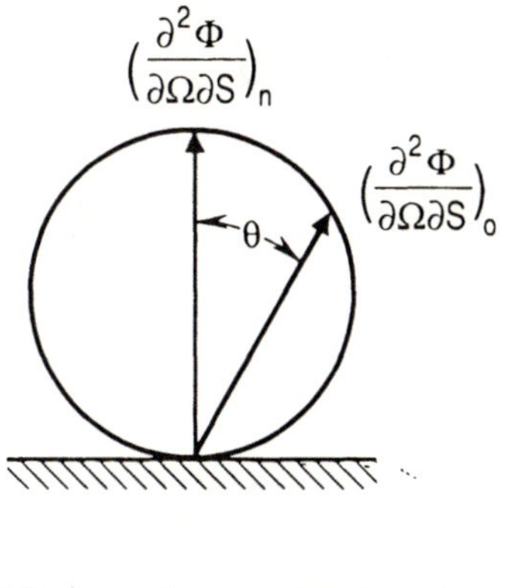

Figure 2.9 Radiant power of a diffusive surface as a function of angle θ.

To calculate the average radiant power per unit solid angle per unit area of a Lambertian or diffusive surface, consider a sphere with unit radius as shown in Figure 2.10. The total radiated power, $d\Phi$, in the annulus $d\theta$ can be calculated with eq. (2.26). If the normal direction to the surface element dS_1 coincides with the Z axis, we can write

$$
\begin{aligned}
d\Phi &= \left(\frac{\partial^2\Phi}{\partial\Omega\partial S}\right)_\theta \cdot 2\pi \sin\theta \, d\theta \, dS_1 \\
&= 2\pi L_n \cos\theta \sin\theta \, d\theta \, dS_1
\end{aligned}
\tag{2.28}
$$

The total radiant power, Φ, from the surface element dS_1, for the upper hemisphere surrounding dS_1, is

$$
\begin{aligned}
\Phi &= \int_0^{\pi/2} 2\pi L_n \sin\theta \cos\theta \, d\theta \, dS_1 \\
&= \pi L_n dS_1 [\sin^2\theta]_0^{\pi/2} = \pi L_n \, dS_1
\end{aligned}
\tag{2.29}
$$

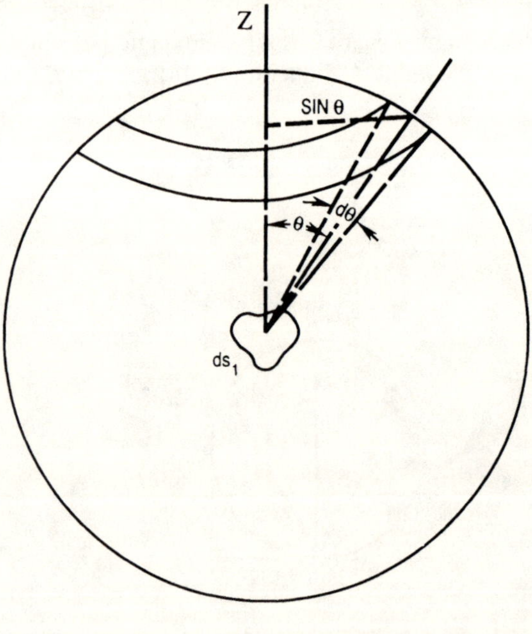

Figure 2.10 Geometry for evaluation of average radiance of a Lambertian surface.

The average radiant power per unit solid angle per unit surface area $(\partial^2\Phi/\partial\Omega\partial S)$ can be obtained by dividing this energy by 2π sr and the area dS_1:

$$\left(\frac{\partial^2\Phi}{\partial\Omega\partial S}\right)_{\text{ave}} = \frac{\pi L_n\, dS_1}{2\pi\, dS_1} = \frac{L_n}{2} \qquad (2.30)$$

In words, the average radiant power per unit solid angle per unit area of a diffusive surface is equal to half its normal radiance.

Most natural surfaces emit and reflect radiation with radiance approximately independent of the viewing angle. In many practical instances, therefore, these surfaces can be considered Lambertian sources of radiation so that their radiance is given by

$$L_n \approx \frac{M}{\pi}\ \text{W-cm}^{-2}\text{-sr}^{-1} \qquad (2.31)$$

where

L_n is the radiance normal to the surface,
M is the radiant exitance of the surface.

2.6 POINT SOURCE

A point source is defined as the limit of a radiation surface when the area of the surface decreases but its radiance increases such that the integral

$$\lim_{\substack{dS_1 \to 0 \\ L_n \to \infty}} \int L_n\, dS_1 \qquad (2.32)$$

exists and is finite. The radiant intensity or the intensity of a point source, designated by I, is the finite limit of this integral. The intensity of a point source is assumed to be isotropic; that is, it radiates with an equal brightness in all directions. It is therefore specified as the power radiated per unit solid angle only, that is, in watts per steradian (W-sr^{-1}).

The radiance, L, of a surface or the radiant intensity of a point source is a function of wavelength, and its value for a very small spectral bandwidth should be specified for that wavelength. The spectral radiance of a surface, therefore, can be expressed in W-cm^{-2}-sr^{-1}-μm^{-1}. The spectral radiant intensity of a point source, on the other hand, can be defined as the radiance in watts per unit solid angle per unit wavelength at a specified wavelength (W-sr^{-1}-μm^{-1}).

2.7 SURFACE IRRADIANCE

The radiant power incident on a unit area of a surface is defined as the irradiance at the surface. The irradiance at a surface due to another radiating surface can be calculated as follows.

Assume a surface element dS_2, located at a distance R from a radiating surface dS_1, is perpendicular to the line AB connecting the two surfaces and subtends a solid angle $d\Omega$ around AB (see Figure 2.11). The differential irradiance dE at the surface S_2 is

$$\mathrm{d}E = \frac{L_n \cos\theta_1 \, dS_1 \, d\Omega}{dS_2} \tag{2.33}$$

where θ_1 is the angle between direction AB and the normal to the surface dS_1. From the figure we see that

$$d\Omega = \frac{dS_2}{R^2} \tag{2.34}$$

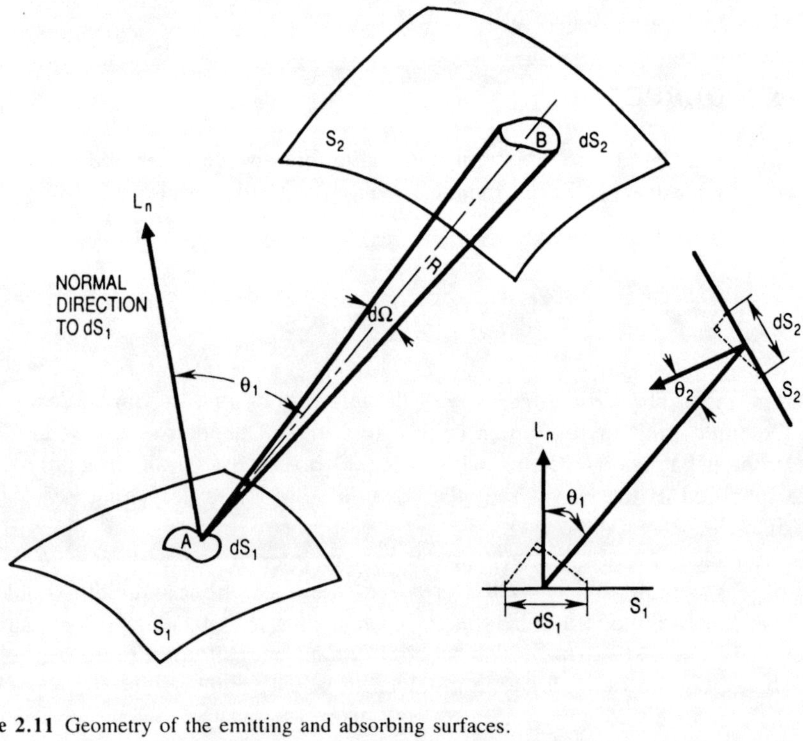

Figure 2.11 Geometry of the emitting and absorbing surfaces.

Inserting eq. (2.34) into eq. (2.33),

$$dE = L_n \cos\theta_1 \frac{dS_1}{R^2} \tag{2.35}$$

If the surface dS_2 is not perpendicular to the line AB, its projection to the surface normal to this direction would be $dS_2 \cos\theta_2$, where θ_2 is the angle between the direction normal to the surface dS_2 and the line AB. In this case, eq. (2.35) becomes

$$dE = L_n \cos\theta_1 \cos\theta_2 \frac{dS_1}{R^2} \tag{2.36}$$

The total irradiance E at the surface dS_2 from all elements of the surface S_1 can be obtained by integrating eq. (2.36) over the surface S_1:

$$E = \int_{S_1} L_n \cos\theta_1 \cos\theta_2 \frac{dS_1}{R^2} \tag{2.37}$$

When S_1 is a plane surface and its largest dimension is small with respect to the range R, then θ_1 can be considered a constant for all elements of the surface S_1. In this case, $dS_1 \cos\theta_1$ is the projected area of the surface dS_1 onto a plane perpendicular to the line AB, and eq. (2.37) reduces to

$$E = L_n \frac{S_1 \cos\theta_1 \cos\theta_2}{R^2} \tag{2.38}$$

As an example, let us evaluate the surface irradiance at the entrance aperture of an optical sensor. The sensor is assumed to have a circular aperture with a diameter D_0 and is located at a distance R from a rectangular-shaped diffusive plane surface S_1, as shown in Figure 2.12. We have

$$\begin{aligned} S_1 &= a \cdot b \\ \theta_2 &= 0° \end{aligned} \tag{2.39}$$

The irradiance at the sensor entrance aperture S_2, according to eq. (2.38) is

$$E = L_n \frac{S_1 \cos\theta_1}{R^2} = L_n \frac{ab \cos\theta_1}{R^2} \tag{2.40}$$

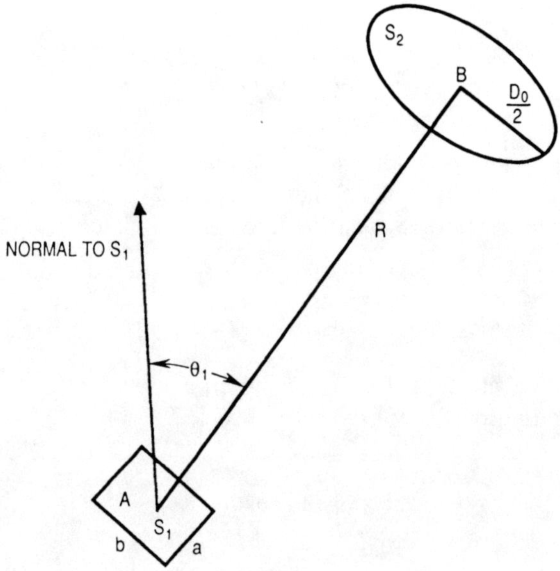

Figure 2.12 Geometry of collecting aperture S_2 and emitting surface S_1.

And the total incident power received by the sensor, P_s, is

$$P_s = EA_0 = L_n \frac{ab \cos\theta_1 \ \pi D_0^2}{4R^2} \tag{2.41}$$

where A_0 is the area of the entrance aperture of the sensor.

As another example, let us consider the irradiance at the entrance aperture of a sensor from a point source target with an intensity of I Watts per unit solid angle, as shown in Figure 2.13. As mentioned before, the radiant intensity of a point source I is

$$\underset{\substack{dS_1 \to 0 \\ L \to \infty}}{\text{limit}} \int L dS_1 \tag{2.42}$$

Using eq. (2.38), and assuming $R \gg r$, the irradiance, E, at the collecting aperture is

$$E = \frac{I}{R^2} \tag{2.43}$$

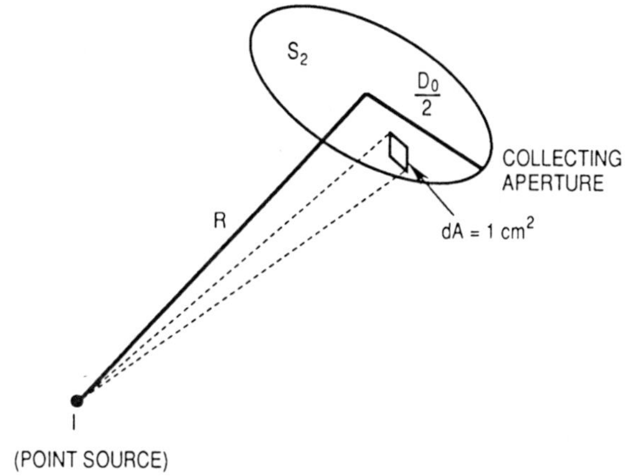

Figure 2.13 Geometry of a point source and a collecting aperture.

The total power received by the sensor entrance aperture, P_s, is

$$P_s = \int_{S_2} E \, dS_2 = \frac{\pi D_0^2}{4} \cdot \frac{I}{R^2} \tag{2.44}$$

Note that when calculating the total power received by the entrance aperture of a sensor, we must take into account for the transmission of the radiant energy through the atmosphere. Atmospheric transmittance, τ_a, defined as the ratio of the transmitted radiant flux through the atmosphere to the incident radiant flux, can be used as an attenuation factor in the calculation of the irradiance at the sensor entrance aperture.

Example

Evaluate the irradiance, E, at the sensor entrance aperture and total incident power at the aperture due to a point source target of 10 W-sr^{-1} intensity located at 100 km from the sensor, as shown in Figure 2.14. The aperture diameter of the sensor is 5 inches. Attenuation of the radiation through the atmosphere is assumed to be 50%.

The irradiance, E, at the sensor entrance aperture is

$$E = \frac{I}{R^2} = \frac{10}{(10^7)^2} = 10^{-13} \text{ W-cm}^{-2}$$

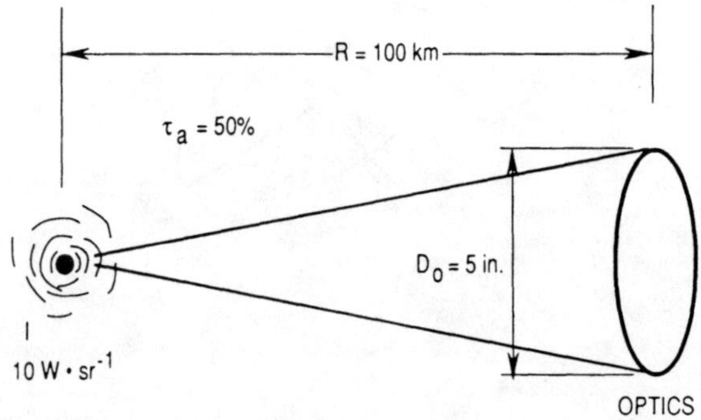

Figure 2.14 Geometry of the example problem with a point source target.

Due to atmospheric attenuation, this value becomes

$$E = \tau_a \frac{I}{R^2} = 0.5(10^{-13}) = 5 \times 10^{-14}\ \text{W-cm}^{-2}$$

The total power received at the sensor entrance aperture is

$$P_s = W A_0 = 5 \times 10^{-14} \cdot \frac{\pi D_0^2}{4} = \frac{(5 \times 10^{-14})(\pi)(5 \times 2.54)^2}{4}$$

$$= 6.33 \times 10^{-12}\ W$$

Example

Evaluate the total incident power collected by the optical sensor from a rectangular object with radiance 4×10^{-4} W-cm^{-2}-sr^{-1} located 2 mi from the sensor, as shown in Figure 2.15. The sensor line of sight makes an angle of 60° with the plane normal to the surface of the object. The area of the surface is 4 ft × 8 ft. The entrance aperture is 2 inches in diameter. An atmospheric transmittance of 70% is assumed.

The irradiance at the sensor entrance aperture is

$$E = \tau_a L_n \frac{S_1 \cos\theta_1}{R^2}$$

$$= (0.7)(4 \times 10^{-4}) \frac{(4)(8) \cos 60°}{(2 \times 5280)^2}$$

$$= 4 \times 10^{-11}\ \text{W-cm}^{-2}$$

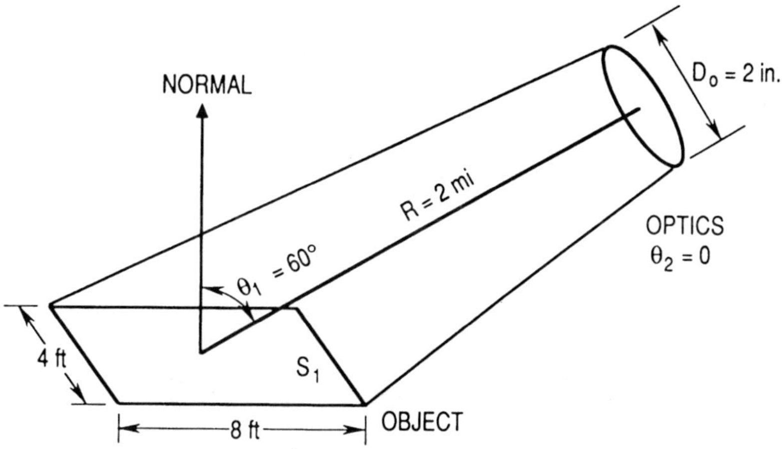

Figure 2.15 Geometry of the example problem with a surface radiating target.

The total power collected at the aperture is

$$P_s = EA_0 = (4 \times 10^{-11}) \frac{\pi D_0^2}{4} = (4 \times 10^{-11}) \frac{(\pi)(5 \times 2.54)^2}{4}$$

$$= 5.06 \times 10^{-9} \text{ W}$$

2.8 WAVELENGTH DEPENDENCE OF SIGNAL POWER

In the preceding example and discussion, we assumed that the target radiated power incident on the receiving aperture was constant over the wavelength region of interest. Variations in target radiated power, atmospheric transmission, optical transmission, and detector response as a function of wavelength, can be accounted for by integrating the incident energy over appropriate wavelengths. The total incident power on the detector then becomes

$$P_s = \frac{A_0}{R^2} S_1 \int_0^{\infty} I(\lambda)\, \tau_a(\lambda)\, K_0(\lambda)\, R(\lambda)\, d\lambda \qquad (2.45)$$

where

P_s is the incident power on the detector,
A_0 is the receiving aperture area,

R	is the sensor-target range,
S_1	is the target area,
$I(\lambda)$	is the target spectral intensity,
$\tau_a(\lambda)$	is the atmospheric spectral transmission,
$K_0(\lambda)$	is the optical spectral transmission,
$R(\lambda)$	is the detector spectral response,
λ	is the wavelength.

Note that having the values of these parameters as a function of wavelength the preceding integral can be evaluated numerically by obtaining the product of these values and finding the area under the resulting curve.

The target signal strength discussed in this chapter travels through space or atmosphere before reaching the receiving sensors. If the signal has to travel through the atmosphere, atmospheric losses should be taken into account, as will be discussed in the next chapter. The resulting signal then enters the electro-optical system and finally the detector for detection. In the detector the signal should compete with detector noise for detection. That is, if the target is in a clear background; otherwise, target background noise will increase the detector noise, making target detection more difficult. Detector and background noise characteristics will be discussed in Chapter 4.

2.9 EXTERNAL RADIATION SOURCES

Optical systems function by sensing the radiation emitted by an object or the radiation reflected from the object being illuminated by an external radiation source. Detection of a jet engine plume by an infrared detector is an example of the first category whereas the visible detection of an area illuminated by the sun is an example of the second category. In both cases these sensors are passive, as they do not generate their own power. Active sensors on the other hand generate their own power and detect targets from the reflected energy of this power. Electro-optical systems may operate using natural and artificial sources of power.

2.9.1 The Sun as a Source

The sun is the most common source of external radiation for optical sensors. The solar spectral radiance is that of a 6,000 K blackbody. Solar irradiance at the Earth's surface is distinctively different from the solar irradiance outside the atmosphere (see Figure 2.16). This is mainly due to scattering and absorption of radiation by atmospheric constituents such as H_2O, CO_2, and O_3.

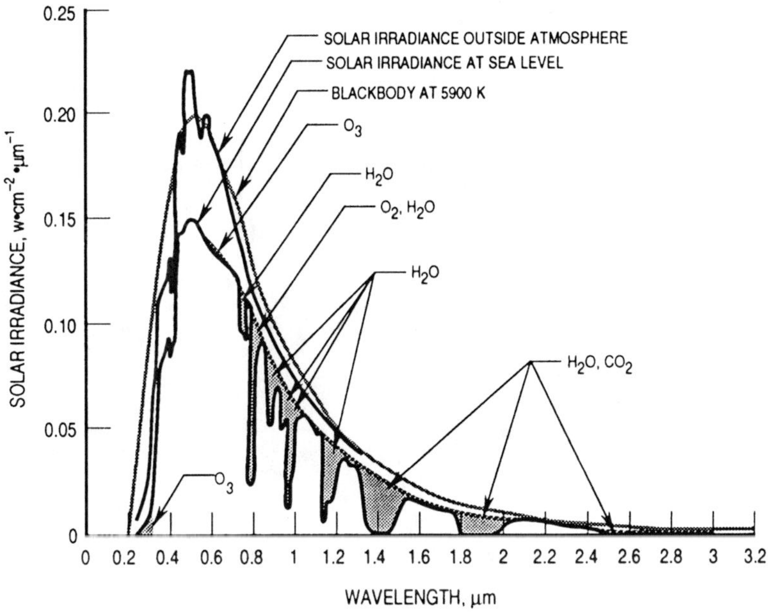

Figure 2.16 Solar spectral irradiance at Earth's surface and outside of Earth's atmosphere (shaded areas indicate absorption at sea level, due to the atmospheric constituents).

2.9.2 Lamps as Sources

Lamps have also been used as radiation sources in many optical system applications. Their spectral radiance is continuous, like a blackbody, but with superimposed line radiation. The tungsten filament incandescent lamp, which operates up to 3,000 K, is the most commonly used lamp. Carbon arc lamps are also used as optical sources. Their spectral radiation approximates that of a 3,800 K blackbody. Gaseous arc discharge lamps are also often used as radiation sources. Xenon lamps are the most common, although argon, krypton, cesium, and mercury lamps have also been used. Figure 2.17 shows the spectral radiance of a 10 kW xenon discharge tube. Input power for the gaseous discharge lamp ranges from 150 W to 30 kW.

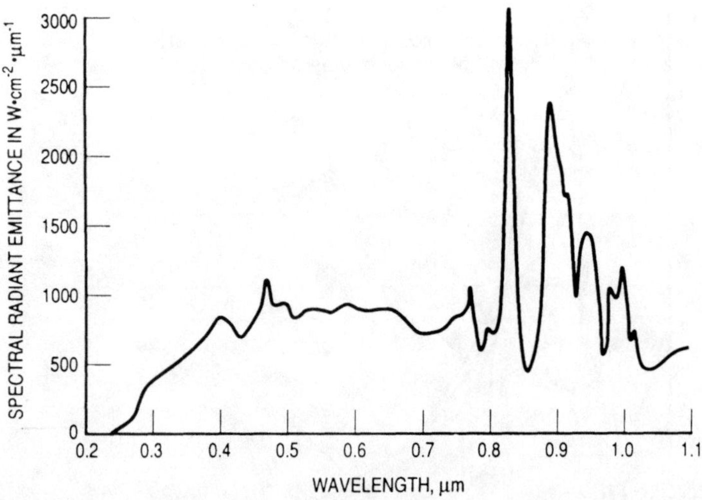

Figure 2.17 Spectral radiant emittance of a 10-kW xenon discharge tube.

2.9.3 Lasers as Sources

Laser and coherent sources are being used more frequently for detection and communication in optical systems. Their high radiation power, narrow spectral band, and coherency are useful in many optical system applications.

Monochrometers can filter incoherent light down to a very narrow spectrum, but only with considerable loss of intensity. The output of the laser, however, is inherently monochromatic and directive as well. Figure 2.18 shows a typical

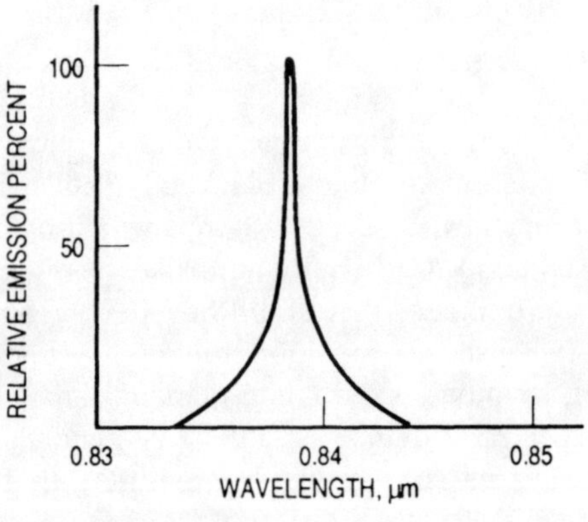

Figure 2.18 A typical spectral radiance of a laser beam.

spectral radiance of a laser source. The laser spectrums, however, are extremely narrow.

2.10 SURFACE REFLECTIVITY

The reflected radiation from an object illuminated by an external radiation source depends on the intensity of the source and the reflectivity of the object surface. This reflected energy can be considered the source of target power for an electro-optical sensor system. This is the case in visible optical sensors, where the reflected light energy is picked up by the sensor.

Surface reflectivity is defined as the fraction of the irradiance of the surface that is reflected. Surface reflectivity is wavelength dependent, it is related to incident and reflected energy as follows:

$$M(\lambda) = E(\lambda)\,\rho(\lambda) \tag{2.46}$$

where $M(\lambda)$ is the reflected radiation, $E(\lambda)$ is the incident energy, and $\rho(\lambda)$ is the surface reflectivity. Figure 2.19 gives the surface reflectivity of commonly encountered metals as a function of wavelength. Similarly, Figure 2.20 gives the spectral reflectance of vegetated and nonvegetated land areas [7]. These reflectivity values may also be used in land use and crop recognition studies using imaging spectrometers.

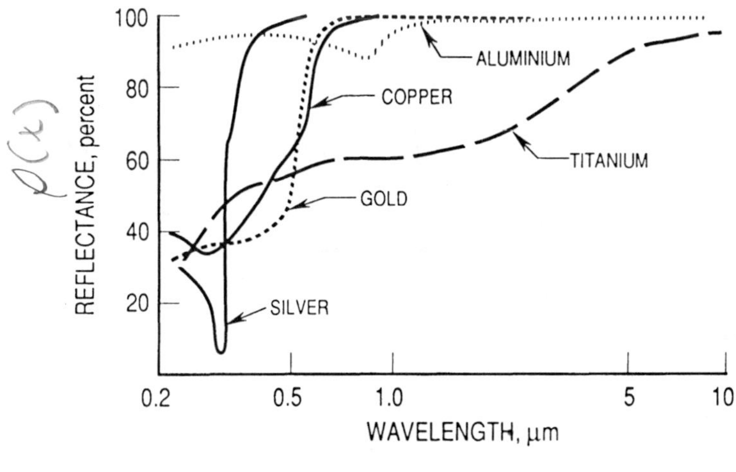

Figure 2.19 Reflectance characteristics of commonly used metals.

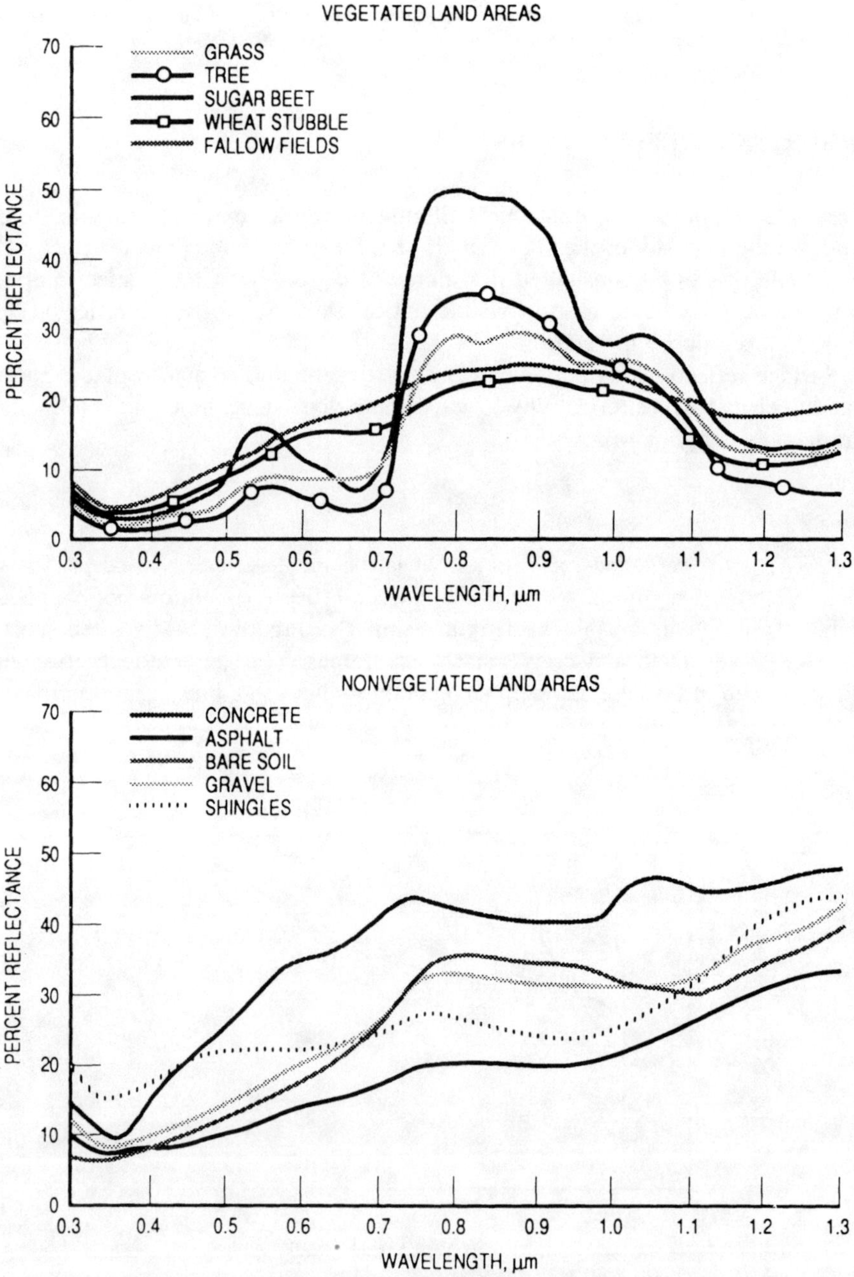

Figure 2.20 Reflectivity characteristics of land cover types.

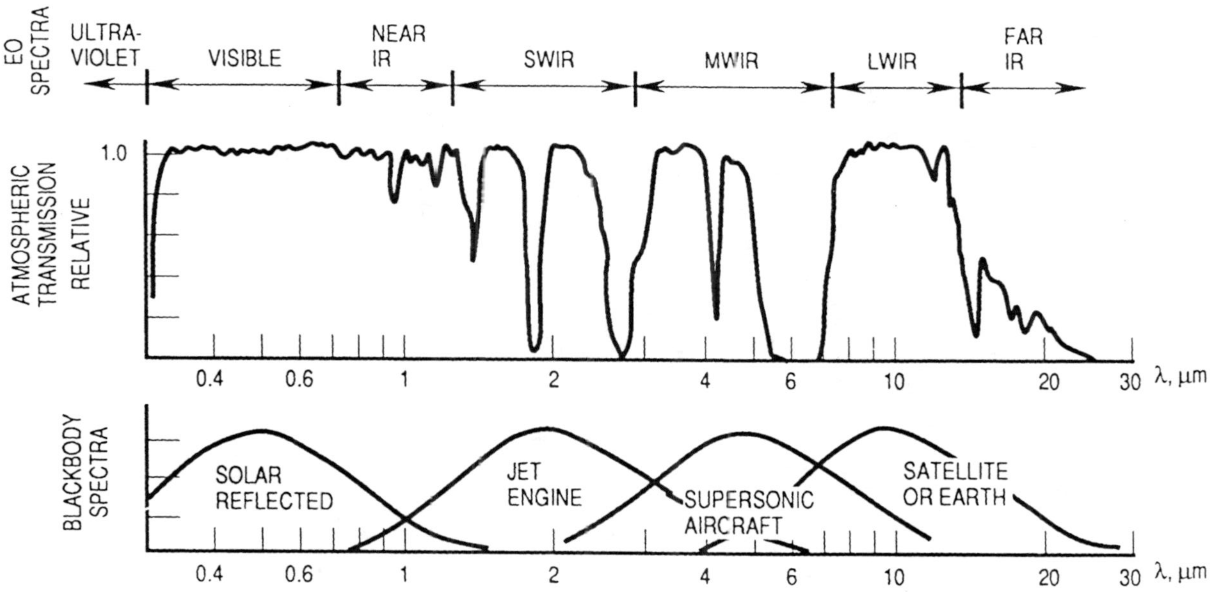

Figure 2.21 Typical spectral return of several EO targets

2.11 TARGET RADIATION SIGNATURE

Target radiation must be detected and discriminated against all background radiation within the field of view of an EO sensor. This task is very difficult, if not impossible, when the target radiation is very similar to the background radiation. Due to similarities, the background may be a source of false targets and clutter. The target radiation features must be known if it is to be distinguished from the background.

Distinction between the target and the background may be observed by a sensor when the temperature, emissivity, or reflectivity of the target is different from the background. In the detection of a terrestrial vehicle, for example, a sensor may detect the temperature differences that arise between the vehicle and the terrain background due to the heating of the vehicle's engine.

For airborne targets such as aircraft or missiles, the radiation from plume emissions, hot metal parts, and aircraft skin heating may be detected against the sky or ground background. Plume emissions consist of hot molecules that radiate large amounts of radiation. Parts of the target may be hot due to heating from the engines, and these may be another source of radiation. For supersonic aircraft, the skin of the aircraft may be heated, producing a large amount of radiation that dominates the overall target signature. Figure 2.21 shows a typical blackbody source spectra of airborne targets. The same figure also shows the atmospheric transmission characteristics as a function of wavelength for reference.

REFERENCES

[1] Seyrafi, K., *Electro Optical System Analysis*, Los Angeles: Electro Optical Research Company, 1973.

[2] Jamieson, J.E., et al., *Infrared Physics and Engineering*, New York: McGraw-Hill, 1963.

[3] Seyrafi, K. (Ed.), *Engineering Design Handbook, Infrared Military Systems*, Washington, DC: U.S. Army Material Command, 1971.

[4] Wolfe, W.L., and G.J. Zissis (Eds.), *The Infrared Handbook*, Ann Arbor: Environmental Research Institute of Michigan, 1985.

[5] Hudson, R.D., *Infrared System Engineering*, New York: John Wiley and Sons, 1969.

[6] Lloyd, J.M., *Thermal Imaging Systems*, New York: Plenum Press, 1975.

[7] McCartney, E.J., *Absorption and Emission by Atmospheric Gases*, New York: John Wiley and Sons, 1983.

Chapter 3
Atmospheric Transmission

3.1 INTRODUCTION

This chapter discusses the essential theories and experimental data for the transmission and absorption of optical radiation through the atmosphere. Because the problems of such absorption and emission affect the performance and design of EO sensors, the material of this chapter serves as a critical link between the source and the receiver of optical radiation and sensor parameters.

The interaction of optical rays with atmospheric gaseous molecules and particles is a very interesting and complex process. Although this subject requires a very specialized knowledge, it is not possible to design or evaluate the performance of an EO sensor without such knowledge.

This chapter is written with the particular intention of presenting the basic theory of interaction of light with atmospheric gases on a solid theoretical ground. Later, a simplified rule for calculating a relatively accurate atmospheric transmission is given. Fortunately, it is possible to incorporate the myriad parameters of atmospheric constituents with variable concentrations into a computer program to evaluate the transmission of any optical beam through a given optical path.

It is instructive, however, to give a brief description of the atmosphere itself before delving into its detailed optical characteristics. The atmosphere can be considered a gaseous blanket that provides a temperature-controlled environment for Earth's inhabitants. Its existence has kept Earth's temperature constant within a relatively small variation. At present Earth is in a radiative thermal equilibrium with space; that is, the average solar radiation received by Earth is equal to the radiant energy escaping from Earth into space. As energy emitted from the Earth depends on the atmospheric temperature, so the temperature adjusts itself until the outflow energy and inflow energy become equal. At present the average global temperature is about 15°C. At the heart of the thermal control process is the absorption and emission of sunlight by the atmospheric gaseous constituents. This occurs mostly in the lower atmosphere.

One can also see the atmosphere as a universal laboratory for observing the effect of light with matter and, in particular, with the gaseous molecules and particles. Observers here are the whole population of the world. The beautiful blue sky, the astonishing color spectrum of sunrise and sunset, the magnificent extension of the rainbow and its colors, and the reflected sunlight from the formation of cloud structure and its mushrooming effects are visible manifestations of the interaction of light and the atmosphere. Whereas this visible phenomenon has been the source of many poems and poetry, in the invisible spectrum of UV and IR, this amazement grows much deeper as the gaseous molecules and particles in the atmosphere provide myriad absorption lines and emission bands.

Behind the lovely scenery exists the equally elegant but more complex reasoning world of quantum mechanics. The interaction of light quanta with atmospheric gaseous molecules and particles can be understood only on the basis of quantum theory. Although the detailed knowledge of the propagation of electromagnetic waves with the atmosphere may be very complex and difficult for the layperson, some optical characteristics of the atmosphere, as it affects the function of life on this planet, are essential knowledge for everyone. The increase in the concentration of CO_2 in recent years may increase the solar absorption of Earth's atmosphere and reduce the radiative heat transfer to space: a process called the *greenhouse effect*. The process may increase the equilibrium global temperature and could alter the biological life on the planet. In the upper atmosphere, another by-product of civilization may alter the preservation of life in a different way. The ozone layer generated in the upper atmosphere by the interaction of solar UV radiation with oxygen molecules serves in turn to block the same deadly solar UV from reaching Earth's surface. In this process, the ozone layer itself becomes destroyed. The equilibrium point in the generation and destruction of this process is only a thin layer of ozone, which serves as the protector of Earth's inhabitants. The ozone concentration is no more than 3 parts per million of the gases in the atmosphere.

UV radiation, if it escapes the ozone layer, is sufficiently energetic that it can break apart important biological molecules, such as DNA. It can also create skin cancer, cataracts, immune deficiencies, and may harm crops and aquatic ecosystems.

At present, some of the by-products of the advanced societies, such as the complex halogen molecules associated with antiperspirants and oven cleaners are potential threats to the ozone layer. Included in this category is the infusion of nitrogen oxides into the stratosphere by high-flying supersonic aircraft.

In 1971 the possibility of a major fleet of supersonic transport planes caused considerable concern that the resulting exhaust of water vapor and nitrogen oxides would have a harmful effect on the ozone layer. Fortunately, the fleet of supersonic planes never materialized but the rising levels of nitrous oxide from combustion and increased use of nitrogen-rich fertilizers continued. Furthermore, the increased use of compounds such as *chlorofluorocarbons* (CFCs) has a damaging effect on ozone producing processes of the upper atmosphere. The CFC gas is part of coolants for refrigerators and air conditioners, propellant for aerosol sprays, agents for producing foam, and cleansers for electronic parts. The chlorine part of this substance, composed of chlorine, fluorine, and carbon, destroys the ozone molecules. Much environmental regulation has to be employed to keep safe the level of UV radiation reaching the ground.

3.2 ATMOSPHERE, TEMPERATURE, AND PRESSURE PROFILES

Earth's atmosphere is a mixture of many gases with varying characteristics of absorption, emission, and scattering of optical radiation. Each of these gaseous quantities varies with altitude, time, and space as a function of geographical region. The problem of quantitative data on the absorption, emission, and scattering of optical radiation requires a knowledge of patterns of variation of atmospheric gaseous constituents. The variation in pressure, temperature, and the concentration of absorbing gas are the main factors in the calculation of the atmospheric path transmission. Among all atmospheric constituents, water vapor and CO_2 are of main concern here. Figure 3.1 illustrates the stratification of Earth's atmosphere [1].

3.3 PRESSURE AND TEMPERATURE

Pressure is the only atmospheric variable whose variation with time is almost always negligible, below 30 km elevation for atmospheric transmission calculations, although it is important for weather predictions. The variation of pressure with altitude, however, is more crucial in determining atmospheric transmission, especially via the vertical and slanted paths. The vertical pressure profile is practically the same for various regions of the globe. The variation of pressures at sea level, on the global scale, is about 980 to 1040 m bar. Figure 3.2(a) shows an average pressure profile. The temperature profile [1] of the atmosphere is shown in Figure 3.2(b) [2]. The atmosphere can be divided into various layers, depending on temperature variation with altitude. The troposphere is the lowest layer of the atmosphere, characterized in general by a decrease in temperature with increasing altitude; nearly all cloud formations occur in this layer. The troposphere extends from Earth's surface to an altitude of 16 to 18 km in the tropical regions and 7 to 10 km in the polar regions. The upper boundary of the troposphere is known as the tropopause, where the temperature stops decreasing and begins increasing. The temperature at the tropopause varies from about 220 K at the poles to about 190 K at the equator.

In the stratosphere, the next layer, which extends to an altitude of approximately 50 km, the temperature rises slowly up to 20 km elevation, and then rises rapidly. The temperature in this layer reaches a maximum of about 270 K. The upper boundary of the stratosphere is known as the stratopause, where the temperature stops increasing.

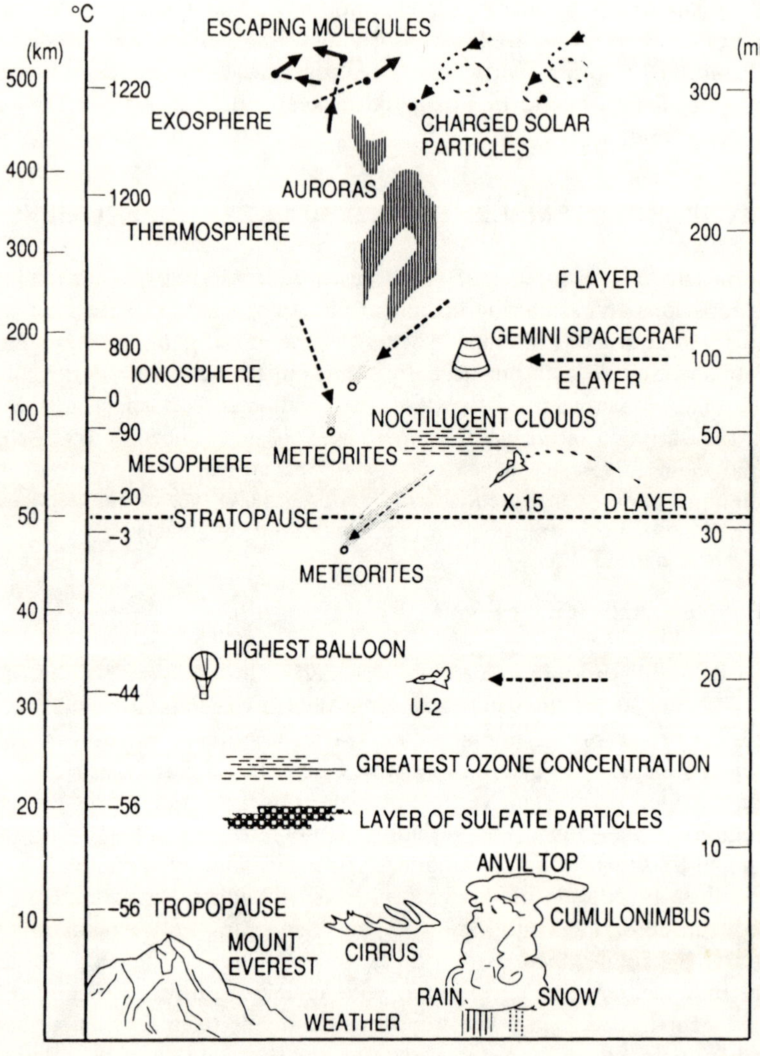

Figure 3.1 Stratification of the atmospheric envelope.

Beyond the stratopause is the mesosphere, characterized by a rapid decrease in temperature. The mesosphere extends from 50 km to about 80 to 90 km in altitude. The mesopause is the upper boundary of the mesosphere and the temperature once more reaches a minimum there.

Above the mesopause is the thermosphere, characterized by a steady increase in temperature. The temperature, increasing with altitude, reaches a maximum of

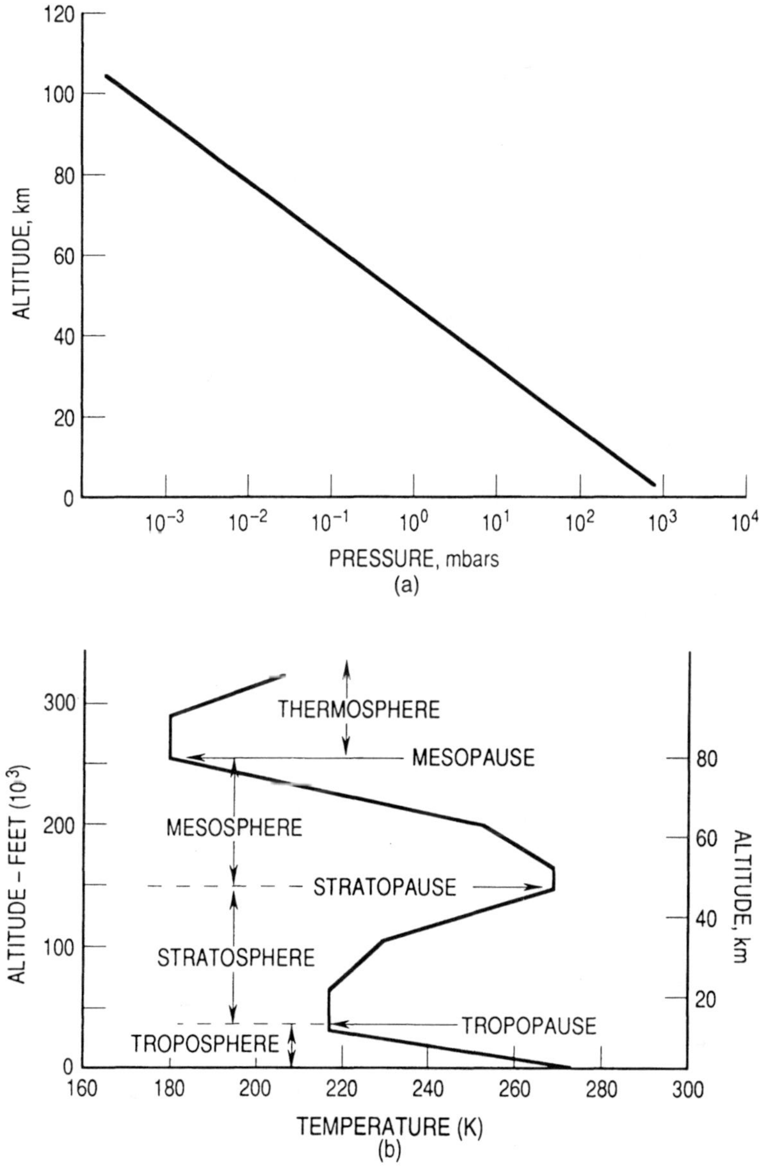

Figure 3.2 Typical average (a) pressure and (b) temperature profiles of the atmosphere.

about 500 K during minimum solar activity at nighttime and about 1,750 K during maximum solar activity in the daytime. The altitude where the temperature stops increasing is known as the *thermopause*.

The region below the stratospause deserves special attention because it contains the majority of the elements that affect the transmission of optical radiation; that is, absorbing and scattering molecules as well as aerosols.

3.4 ATMOSPHERIC CONSTITUENTS

The gaseous contents in the clear atmosphere, and their concentration are indicated in Table 3.1. The major constituents of the atmosphere up to 90 km are molecules of N_2 and O_2. The next highest concentration (by far less than N_2 and O_2) are argon, water vapor, and carbon dioxide.

Table 3.1
Composition of Atmospheric Constituent Gases

Constituent Gas	*Content (% by volume)*
H_2	78.084
O_2	20.9476
Ar	0.934
CO_2	3×10^{-2}
H_2O	10^{-5} to 10^{-2}
Ne	1.81×10^{-3}
He	5.2×10^{-4}
CH_4	2×10^{-4}
Kr	1.14×10^{-4}
H_2	5×10^{-5}
N_2O	$\approx 5 \times 10^{-5}$
CO	$\approx 7 \times 10^{-6}$
O_3	0 to 7×10^{-6}
NO_2	
NO	0 to 2×10^{-6}

Water vapor has a concentration between $10^{-5}\%$ and $10^{-2}\%$ per volume and extends from Earth's surface to the stratosphere (above 25 km). The concentration of water vapor in the atmosphere is subject to a large fluctuation, depending on the altitude, season, time of day, geographic location, meteorological condition, and so forth. Because of this fluctuation, the accurate value of the water vapor in the atmosphere, at any instant of time, is usually not known. However, an approximate value is generally used in the calculation. Actually, water vapor concentration is measured globally, and it is better known than some minor gases.

Carbon dioxide is more uniformly distributed, with a concentration of approximately $3 \times 10^{-2}\%$ per volume. Its value is larger over industrial cities and vegetation fields than over the deserts or oceans.

Ozone, O_3, has a relatively low concentration, but it does have a significant effect on the radiation absorption. A typical ozone concentration at 20 to 30 km elevation is 10^{-4} to $3 \times 10^{-4}\%$ per volume (1 to 3 parts per million), with a much lower concentration at lower altitudes.

Other gaseous constituents responsible for the visible and IR absorption are methane, with a concentration of $(1–2) \times 10^{-4}\%$ per volume, nitrous oxide (N_2O), with a concentration of $5 \times 10^{-5}\%$, and carbon monoxide, with a concentration of about $7 \times 10^{-6}\%$. These gases are of relatively minor importance in the determination of the optical properties of the atmosphere, although they may be more important for laser beam propagation.

3.5 OPTICAL PROPERTIES OF THE ATMOSPHERE

Until a few decades ago, interest and knowledge of the atmospheric phenomena were primarily limited to meteorology, heat transfer, and the physics of the atmosphere. Recent interest in transmission of optical radiation through the atmosphere, such as in optical detection, laser propagation, and remote sensing devices, has generated new incentives in the study of the optical properties of the atmosphere.

Within terrestrial applications, the optical signal radiated from an object must pass through Earth's atmosphere before reaching a receiving aperture. (We exclude here those systems used from a space platform for detection of celestial or exoatmospheric objects.) The effect of the atmosphere on the intensity and spectral signature of the signal is very significant and requires considerable study for a proper treatment of optical propagation problems. This effect is always dominated by the attenuation of the incident radiation by the gases and particles in the atmosphere and by the turbulence and scintillation phenomena of the atmosphere. Considerable efforts have been made during the past three decades to evaluate atmospheric effects on the wide spectrum of electromagnetic radiation, including ultraviolet, infrared, and visible radiation.

The attenuation of optical radiation due to the atmosphere arises from the individual or collective effects of the following phenomena:

1. Molecular absorption,
2. Molecular scattering,
3. Aerosol absorption,
4. Aerosol scattering.

The two main causes of attenuation, however, are molecular absorption by several minor constituents of the atmosphere and scattering due to atmospheric molecules on the presence of particles in the atmosphere (aerosols). Molecular absorption, as will be discussed later, occurs mainly in several more or less narrow absorption bands and is due to the ability of certain molecules to go from one state of vibrational rotation to another, thereby absorbing or emitting a photon. Figure 3.3 shows the typical sunlight spectrum as it passes through Earth's atmosphere [3]. The distortion of the spectral intensity profile and some of the absorption bands are clearly evident. Scattering, on the other hand, causes redistribution of the incident radiation energy into all directions of propagation; therefore, the radiant energy in the original direction diminishes.

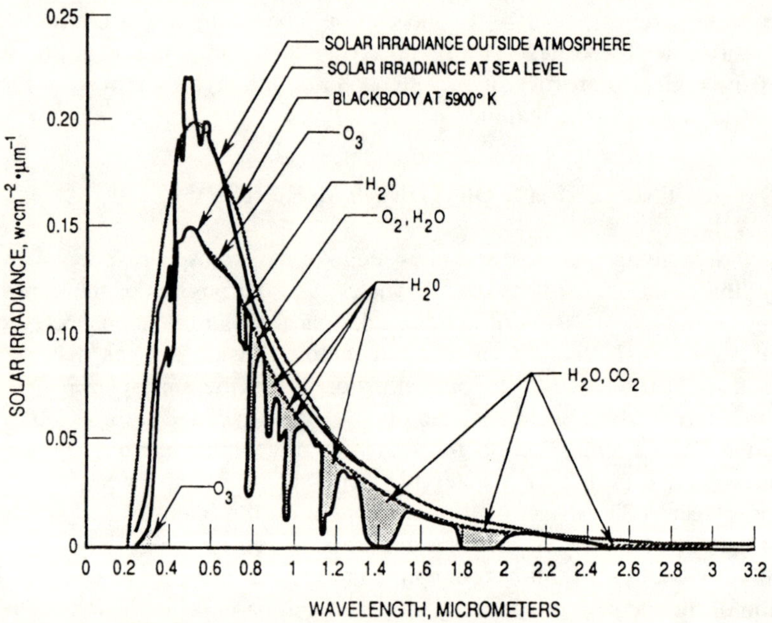

Figure 3.3 Solar spectral irradiance at Earth's surface and outside Earth's atmosphere (shaded areas indicate absorption, at sea level, due to the atmospheric constituents).

The optical properties of the atmosphere in different regions of the spectrum could originate from distinctly different mechanisms, depending on the incident photon energy and the dominant atmospheric constituents in that spectrum. As was stated earlier, the energy of a photon is an increasing function of its wave number, $\bar{\nu}$, defined as the inverse of wavelengths; that is, $\bar{\nu} = \lambda^{-1}$.

Radiation absorption due to molecules in a gaseous medium can be divided into two processes:

1. Atomic absorption is due to the transition of electrons in an atom, and usually requires a photon with an energy of a few electron volts. It generally takes place in the visible or ultraviolet region of the spectrum.

2. Molecular absorption is due to the transition between electronic, vibrational, or rotational energy states of a gas molecule or due to a combination of all of these.

The atomic absorption spectrum consists of many discrete lines, usually widely separated. The molecular absorption spectrum usually consists of many bands, each band composed of many closely spaced discrete lines. The absorption band spectrum may arise from the transition between different vibrational energy levels of the same electronic state, whereas the discrete lines within the absorption band are due to the rotation of the molecules. The electronic absorption band spectrum is due to transition between different vibration-rotation levels of two electronic states.

Figure 3.4 shows the absorption for various gaseous constituents within the atmosphere [3]. Each curve is marked by regions of low absorption known as *windows* and by regions of high absorption where the atmosphere is virtually opaque. Each gas has an absorption signature as a function of its wavelength. The aggregate absorption due to all the gases is depicted by the composite curve at the bottom.

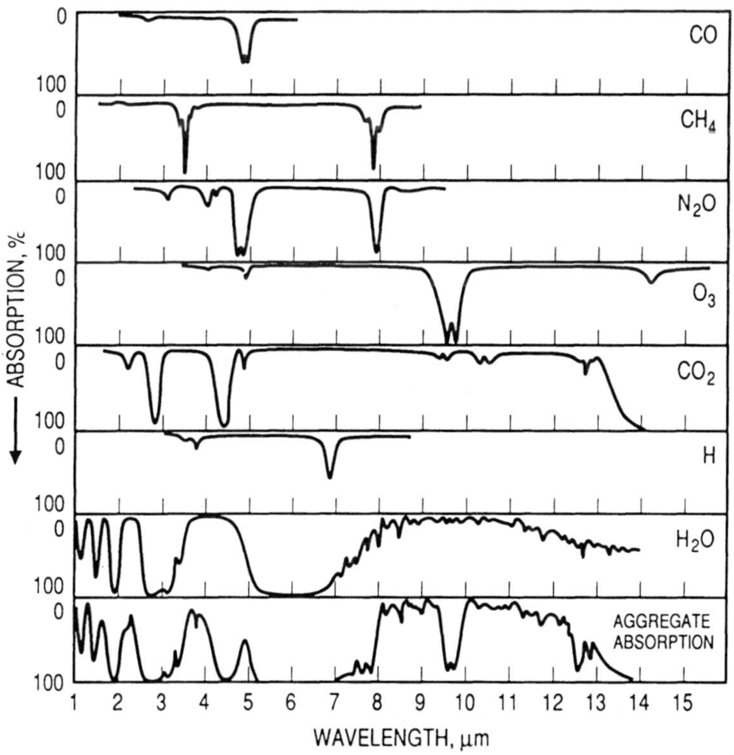

Figure 3.4 Infrared absorption by atmospheric gases.

This is the absorption spectrum revealed by observing the sun from Earth's surface with the sighting path thereby extending completely through the atmosphere. When the sun is at a low elevation angle, the resulting long optical path makes it possible to measure the absorptions of trace gases—often existing as ions and free atoms in the upper air—that otherwise would remain undiscovered. Such observations yield much information about constituents at high altitudes where direct sampling or measurement is not feasible.

Figure 3.5 represents the result of experiments that give the atmospheric transmission at sea level from a wavelength of 0.3 μm to 30 μm. These atmospheric transmission curves consist of successively high transmission regions (atmospheric windows) and low transmission bands (absorption bands).

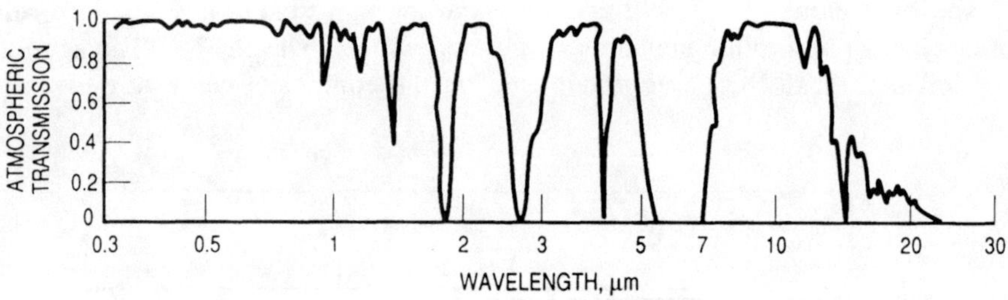

Figure 3.5 Transmission characteristics of electromagnetic radiation through the clear atmosphere at micrometer wavelengths.

3.6 RAYLEIGH SCATTERING

When the atmosphere does not contain large-size particles, but only the molecules of its gaseous constituents, the scattering phenomena is dominated primarily by these molecules. The intensity of the scattering depends on the number and index of refraction of these gaseous molecules and on the scattering angle, wavelength, and polarization of the incident radiation.

The attenuation of a light beam due to Rayleigh scattering is expressed by

$$I = I_0 \, e^{-\sigma_m R} \tag{3.1}$$

where

I_0 is the magnitude of the original radiation,
I is the magnitude of the scattered radiation,
σ_m is the Rayleigh scattering coefficient,
R is the path length.

The Rayleigh scattering coefficient σ_m is the product of two other parameters:

$$\sigma_m = \sigma_r \cdot N \tag{3.2}$$

where

σ_r is the Rayleigh scattering cross section,
N is the molecular number density.

The Rayleigh scattering cross section, σ_r, is defined as the cross section of an incident wave, so that the total power scattered by a gaseous molecule in all directions is equal to the power flowing across the cross section [1]. It is given by the expression

$$\sigma_r = \frac{8\pi^3(n^2 - 1)^2}{3N^2 \lambda^4} \tag{3.3}$$

where

n is the index of refraction of the gaseous medium,
N is the molecular number density.

Note that the scattering cross section σ_r is a function of wavelength and proportional to λ^{-4}. It therefore dominates more in the ultraviolet and visible regions than in the infrared region. The formula also indicates that σ_r is a function of the index of refraction of air and molecular number density. A typical value of σ_r for sea level and good weather is about 0.0172 km^{-1} for $\lambda = 0.5$ μm and about 4×10^{-6} km^{-1} for $\lambda = 4$ μm.

Another useful factor in calculating the Rayleigh scattering is the optical thickness T_r, defined as

$$T_r = \int_0^R \sigma_m \, \mathrm{d}R \tag{3.4}$$

where R is the path length.

The optical transmission τ_r is given by

$$\tau_r = \mathrm{e}^{-T_r} \tag{3.5}$$

Because the molecular number density is altitude dependent, σ_m is a function of altitude and τ_r should be calculated for the desired altitude. Table 3.2 gives the values of σ_m and σ_a (aerosol attenuation coefficient) at sea level at various wavelengths.

Table 3.2
Rayleigh and Aerosol Attenuation Coefficients at Sea Level

λ	$\sigma_m(0)$	$\sigma_a(0)$
0.27	2.282×10^{-1}	0.29
0.28	1.948×10^{-1}	0.27
0.30	1.446×10^{-1}	0.26
0.32	1.098×10^{-1}	0.25
0.34	8.494×10^{-2}	0.24
0.36	6.680×10^{-2}	0.24
0.38	5.327×10^{-2}	0.23
0.40	4.303×10^{-2}	0.20
0.45	2.644×10^{-2}	0.180
0.50	1.716×10^{-2}	0.167
0.55	1.162×10^{-2}	0.158
0.60	8.157×10^{-3}	0.150
0.65	5.893×10^{-3}	0.142
0.70	4.364×10^{-3}	0.135
0.80	2.545×10^{-3}	0.127
0.90	1.583×10^{-3}	0.120
1.06	8.458×10^{-4}	0.113
1.26	4.076×10^{-4}	0.108
1.67	1.327×10^{-4}	0.098
2.17	4.586×10^{-5}	0.085
3.50	6.830×10^{-6}	0.070
4.00	4.002×10^{-6}	0.063

Equation (3.4) for a path length R can be approximated by

$$T_r = \sigma_m \cdot R \tag{3.6}$$

and eq. (3.5) can be written as

$$\tau_r = e^{-\sigma_m \cdot R} \tag{3.7}$$

As an example, consider the attenuation over a 5 km horizontal path at sea level for a radiation wavelength of 0.5 μm.

First, we obtain the coefficient $\sigma_m(0)$ from Table 3.2:

$$\sigma_m(0) = 1.7 \times 10^{-2}\ \text{km}^{-1}$$

Then we find τ_r as

$$\tau_r = e^{-1.7 \times 10^{-2} \times 5} = 0.92$$

3.7 AEROSOL SCATTERING

Aerosol scattering, or Mie scattering, is concerned with the radiation effect on spherical particles of various sizes. The transmitted beam energy here, similar to Rayleigh scattering, is expressed by

$$I = I_0 \, \mathrm{e}^{-\sigma_a \cdot R} \tag{3.8}$$

where I, I_0, and R are as defined previously, and σ_a is the aerosol attenuation coefficient, which is a function of aerosol density.

Aerosol size distribution does not change much with altitude, whereas the aerosol density does. The relationship for aerosol attenuation coefficient, $\sigma_a(h)$ at altitude h is given by

$$\sigma_a(h) = \frac{M(h)}{M(0)} \sigma_a(0) \tag{3.9}$$

where $M(h)$ denotes the aerosol number density at height h and $\sigma_a(0)$ is the aerosol attenuation coefficient at sea level.

Aerosol density at sea level is approximately 200 cm^{-3}. As altitude increases, the aerosol density decreases with a slight increase at around 15 to 25 km. Figure 3.6 shows a typical aerosol number density versus altitude.

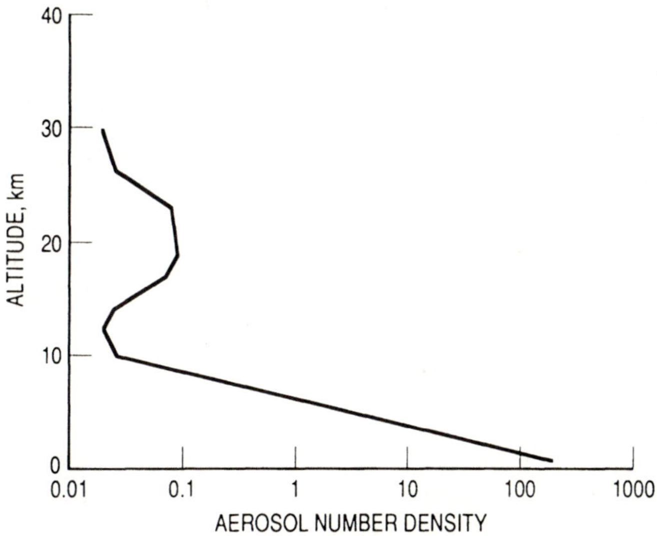

Figure 3.6 Representative profile, aerosol number density versus altitude.

Haze

Haze is generally composed of particles up to a few micrometers in size. These particles consist of dust, salt, and organic residues sometimes surrounded by water droplets. The extinction coefficient of these particles falls off approximately at $\lambda^{0.7}$, or close to Mie scattering.

Transmittance of haze in the infrared region is much higher than in the visible region. There is very little absorption in most hazes, and the transmission is dominated by scattering.

For example, in the 8 μm to 14 μm atmospheric window, the absorption coefficient for hazes whose largest drop size is 2 μm, is smaller than 0.01 km^{-1}. The largest extinction coefficient is 0.08 km^{-1}, which corresponds to a haze in which the largest drops are of 4.48 μm diameter. A 20-km path at sea level, for example, has a transmission of about 20%.

3.8 NUMERICAL CALCULATION

Due to the large number of parameters involved in optical transmission through the atmosphere, numerical calculations of atmospheric transmission is an unavoidable process. The aim of the numerical calculation is to predict with a high degree of accuracy the transmittance through the atmosphere, given a path, atmospheric conditions, wavelength, and a set of measured or predicted meteorological parameters. The computer codes applied here are transmission models or computational techniques implemented in computer programs that generate the transmittance or a transmittance spectrum, given the necessary input by the user.

Two numerical methods for atmospheric transmission widely in use are HITRAN and LOWTRAN, developed at the Air Force Geophysics Laboratory (AFGL). HITRAN contains a molecular database for high-resolution transmission calculations, whereas LOWTRAN contains a low-resolution transmission code. As new measurements become available and better understanding of the extinction process are reached, HITRAN and LOWTRAN are modified to give more accurate results and expand their capabilities to a wider range of atmospheric conditions and a broader spectral region.

3.8.1 HITRAN Atmospheric Transmission Code

HITRAN provides a high-resolution spectrum of the atmospheric transmission, making it ideal for laser transmittance calculations. Its spatial resolution is about 0.5 cm^{-1}. HITRAN is not a transmission model, but a molecular data compilation to be used with line-by-line transmittance codes.

The original HITRAN database contained parameters of seven atmospheric molecules: H_2O, CO_2, O_3, N_2O, CO, CH_4, and O_2. More recently, a second database on minor gases (trace gases) has been created. These two databases were then combined for a new HITRAN database that provided more accurate predictions. Also, new measurements and error corrections continually update and modify the HITRAN database, improving the HITRAN results [4].

From detailed knowledge of spectroscopic line parameters of all molecular absorption lines within a spectral region of interest, a line-by-line monochromatic calculation method can be used to obtain a high-resolution spectrum of the atmospheric transmittance. The computations in this method involve the following:

1. A delta function is provided at each absorption line by the molecular database.
2. An appropriate absorption line shape is applied to each absorption line to account for broadening effects.
3. An atmospheric model is introduced.
4. Beer's law is used to compute a closely spaced network of monochromatic transmittances to generate a high-resolution spectrum.

The spectral line shapes used in the HITRAN calculations, which effectively broaden the spectral line to realistic values, are the Lorentz line shape in the lower atmosphere for pressures larger than 10 mbar, a Voight line shape in the pressure region from 10 mbar to 0.1 mbar, and a doppler line shape at higher altitudes.

The transmittance values are calculated at small frequency increments, producing a high-resolution transmission spectrum, such as that of Figure 3.7.

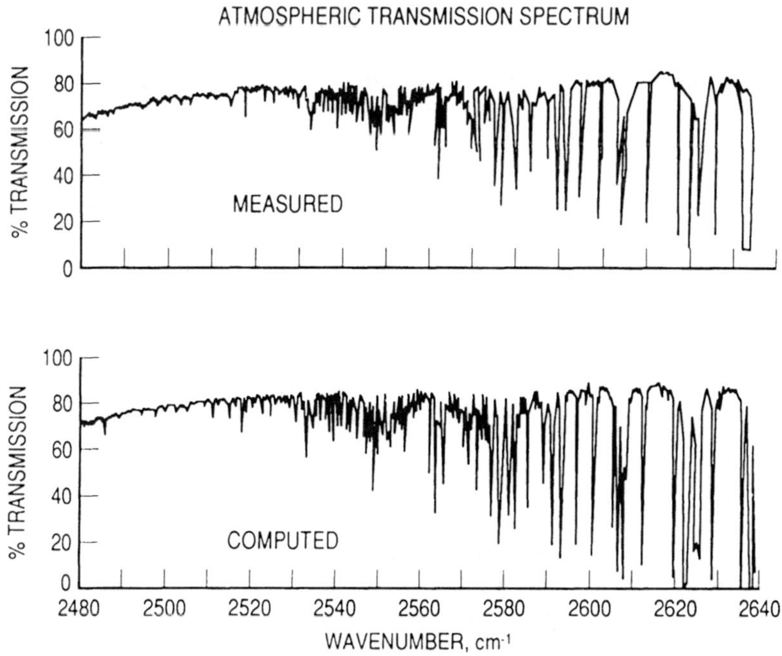

Figure 3.7 Comparison of HITRAN computed high-resolution spectrum with measurement.

Measurements of the transmission values in comparison to the HITRAN results indicate a fairly good agreement.

The computational requirements of the transmittance calculations just described, using the HITRAN database, are enormous. Such computations would require large amounts of computer time and are not practical. Therefore, a computer code called FASCODE [5] which stands for fast atmospheric signature code, has been developed to speed up the HITRAN calculations. FASCODE is an accelerated, line-by-line calculation method that uses a mathematical procedure to reduce the computation time. It uses an algorithm for the accelerated convolution of line shape functions with spectral lines. This method samples the spectral lines in an optimal manner, so that computation is carried out efficiently. It involves separation of the traditional line shape into several subfunctions, each defined for optimal sampling. The sum of these subfunctions form the original line shape function.

Another line-by-line computer code that uses the HITRAN database is the LASER code. It is capable of predicting laser transmission results by generating monochromatic, laser extinction coefficients at selected frequencies [6].

3.8.2 LOWTRAN Atmospheric Transmission Code

LOWTRAN is a computer code that estimates atmospheric transmittance at low resolutions over a wide spectral interval. This transmission model has great computer efficiency, flexibility, and reasonable accuracy for broadband systems applications. It covers the spectral region from 0.25 μm to 28.5 μm at a spatial resolution of 20 cm^{-1} [4].

In LOWTRAN, the total atmospheric transmittance at a given wavenumber is averaged over a 20 cm^{-1} interval. The average transmittance due to molecular line absorption, molecular scattering (Rayleigh scattering), molecular continuum absorption, and aerosol extinction (absorption and scattering) must be determined separately, then multiplied together to give a total average transmittance. That is,

$$\begin{aligned}\bar{\tau}_{\Delta\nu}\,(\text{total}) = {} & \bar{\tau}_{\Delta\nu}\,(\text{molecular line absorption}) \\ & \times \bar{\tau}_{\Delta\nu}\,(\text{molecular scattering}) \\ & \times \bar{\tau}_{\Delta\nu}\,(\text{molecular continuum absorption}) \\ & \times \bar{\tau}_{\Delta\nu}\,(\text{aerosol extinction})\end{aligned} \tag{3.10}$$

The LOWTRAN method started as a graphical method using transmission prediction charts developed at the AFGL. A very convenient and easy method can be used with the charts for graphical calculation of the atmospheric transmission. This graphical method, presented in detail by McClatchey [6], is summarized in the following paragraphs.

A series of five model atmospheres were defined in terms of pressure, temperature, air density, water vapor concentration, and ozone concentration, as functions of altitude from 0 to 100 km: *tropical, midlatitude summer* (MS), *midlatitude winter* (MW), *subarctic summer* (SS), and *subarctic winter* (SW). Table 3.3 gives an example of this for midlatitude summer condition. Table 3.4 lists the concentration of the uniformly mixed gases. In addition to these molecular model atmospheres, two aerosol models were defined: (1) 23 km visibility for clear atmosphere, and (2)

Table 3.3

Model Atmospheres Used as a Basis of the Computation of Attmospheric Optical Properties: Midlatitude Summer

Ht. (*km*)	*Pressure* (*mb*)	*Temp.* (*K*)	*Density* (g/m^3)	*Water Vapor* (g/m^3)	*Ozone* (g/m^3)
0	1.013*E* + 03	294.0	1.191*E* + 03	1.4*E* + 01	6.0*E* − 05
1	9.020*E* + 02	290.0	1.080*E* + 03	9.3*E* + 00	6.0*E* − 05
2	8.020*E* + 02	285.0	9.757*E* + 02	5.9*E* + 00	6.0*E* − 05
3	7.100*E* + 02	279.0	8.846*E* + 02	3.3*E* + 00	6.2*E* − 05
4	6.280*E* + 02	273.0	7.998*E* + 02	1.9*E* + 00	6.4*E* − 05
5	5.540*E* + 02	267.0	7.211*E* + 02	1.0*E* + 00	6.6*E* − 05
6	4.870*E* + 02	261.0	6.487*E* + 02	6.1*E* − 01	6.9*E* − 05
7	4.260*E* + 02	255.0	5.830*E* + 02	3.7*E* − 01	7.5*E* − 05
8	3.720*E* + 02	248.0	5.225*E* + 02	2.1*E* − 01	7.9*E* − 05
9	3.240*E* + 02	242.0	4.669*E* + 02	1.2*E* − 01	8.6*E* − 05
10	2.810*E* + 02	235.0	4.159*E* + 02	6.4*E* − 02	9.0*E* − 05
11	2.430*E* + 02	229.0	3.693*E* + 02	2.2*E* − 02	1.1*E* − 04
12	2.090*E* + 02	222.0	3.269*E* + 02	6.0*E* − 03	1.2*E* − 04
13	1.790*E* + 02	216.0	2.882*E* + 02	1.8*E* − 03	1.5*E* − 04
14	1.530*E* + 02	216.0	2.464*E* + 02	1.0*E* − 03	1.8*E* − 04
15	1.300*E* + 02	216.0	2.104*E* + 02	7.6*E* − 04	1.9*E* − 04
16	1.110*E* + 02	216.0	1.797*E* + 02	6.4*E* − 04	2.1*E* − 04
17	9.500*E* + 01	216.0	1.535*E* + 02	5.6*E* − 04	2.4*E* − 04
18	8.120*E* + 01	216.0	1.305*E* + 02	5.0*E* − 04	2.8*E* − 04
19	6.950*E* + 01	217.0	1.110*E* + 02	4.9*E* − 04	3.2*E* − 04
20	5.950*E* + 01	218.0	9.453*E* + 01	4.5*E* − 04	3.4*E* − 04
21	5.100*E* + 01	219.0	8.056*E* + 01	5.1*E* − 04	3.6*E* − 04
22	4.370*E* + 01	220.0	6.872*E* + 01	5.1*E* − 04	3.6*E* − 04
23	3.760*E* + 01	222.0	5.867*E* + 01	5.4*E* − 04	3.4*E* − 04
24	3.220*E* + 01	223.0	5.014*E* + 01	6.0*E* − 04	3.2*E* − 04
25	2.770*E* + 01	224.0	4.288*E* + 01	6.7*E* − 04	3.0*E* − 04
30	1.320*E* + 01	234.0	1.322*E* + 01	3.6*E* − 04	2.0*E* − 04
35	6.520*E* + 00	245.0	6.519*E* + 00	1.1*E* − 04	9.2*E* − 05
40	3.330*E* + 00	258.0	3.330*E* + 00	4.3*E* − 05	4.1*E* − 05
45	1.760*E* + 00	270.0	1.757*E* + 00	1.9*E* − 05	1.3*E* − 05
50	9.510*E* − 01	276.0	9.512*E* − 01	6.3*E* − 06	4.3*E* − 06
70	6.710*E* − 02	218.0	6.706*E* − 02	1.4*E* − 07	8.6*E* − 08
100	3.000*E* − 04	210.0	5.000*E* − 04	1.0*E* − 09	4.3*E* − 11

Table 3.4
Concentrations of Uniformly Mixed Gases

Constituent	*Molecular Weight*	*ppm by Volume*
Air	28.97	10^6
CO_2	44	330
N_2O	44	0.28
CO	28	0.075
CH_4	16	1.6
O_2	32	2.095×10^5

5 km visibility for hazy atmosphere as given in Table 3.5. These tables present the basis of calculating the atmospheric transmittance, provided that the wavelength, the path length, and the nature of the path length are known, and that one of the model atmospheres can be accepted for a particular application.

The calculations of the average transmittance in the LOWTRAN method are accomplished by a single-parameter band model of the form

$$\bar{\tau}_{\Delta\bar{\nu}} = f[C(\bar{\nu})\, \Delta L\, P^n] \tag{3.11}$$

where

- $\bar{\tau}_{\Delta\bar{\nu}}$ is the average transmittance over the wavenumber interval $\Delta\bar{\nu}$,
- f denotes a function,
- $C(\bar{\nu})$ is the LOWTRAN wavenumber dependent absorption coefficient,
- ΔL is the effective path length (or absorber concentration),
- P is pressure,
- n is an empirical constant.

The function f and the parameter n were determined empirically using laboratory transmittance measurements and transmittance calculations based on the HITRAN database. The f for water vapor and the combined contributions of the uniformly mixed gases (CO_2, N_2O, CH_4, CO, O_2, N_2) were found to be about the same, but n differed between the two. Mean values for n of H_2O was found to be 0.9 and of the uniformly mixed gases to be 0.75. The parameter n of ozone was determined to be 0.4.

Setting $n = 1$ and $n = 0$ in eq. (3.11) gives a "strong line approximation" and a "weak line approximation," respectively, which are terms common to most band models.

The transmittance $\bar{\tau}$ was plotted against log (ΔLP^n) for each wavenumber, then the best mean curve was determined from the superposition of the plotted curves. The resulting curves are the empirical transmittance functions plotted as a function

Table 3.5
Aerosol Models: Vertical Distributions for "Clear" and "Hazy" Atmospheres

Altitude (km)	Particle Density, N (Particles per cm³) 23 km Visibility Clear	5 km Visibility Hazy
0	2.828E + 03	1.378E + 04
1	1.244E + 03	5.030E + 03
2	5.371E + 02	1.844E + 03
3	2.256E + 02	6.731E + 02
4	1.192E + 02	2.453E + 02
5	8.987E + 01	8.987E + 01
6	6.337E + 01	6.337E + 01
7	5.890E + 01	5.890E + 01
8	6.069E + 01	6.069E + 01
9	5.818E + 01	5.818E + 01
10	5.675E + 01	5.675E + 01
11	5.317E + 01	5.317E + 01
12	5.585E + 01	5.585E + 01
13	5.156E + 01	5.156E + 01
14	5.048E + 01	5.048E + 01
15	4.744E + 01	4.744E + 01
16	4.511E + 01	4.511E + 01
17	4.458E + 01	4.458E + 01
18	4.314E + 01	4.314E + 01
19	3.634E + 01	3.634E + 01
20	2.667E + 01	2.667E + 01
21	1.933E + 01	1.933E + 01
22	1.455E + 01	1.455E + 01
23	1.113E + 01	1.113E + 01
24	8.826E + 00	8.826E + 00
25	7.429E + 00	7.429E + 00
30	2.238E + 00	2.238E + 00
35	5.890E − 01	5.890E − 01
40	1.550E − 01	1.550E − 01
45	4.082E − 02	4.082E − 02
50	1.078E − 02	1.078E − 02
70	5.550E − 05	5.550E − 05
100	1.969E − 08	1.969E − 08

of $\bar{\nu}$ on one scale and as a function of transmittance and a scaling factor (log ΔLP^n), on the other scale. These are called *prediction charts*. An example of it for ozone transmission in the wavelength 3 to 6.4 μm is given in Figure 3.8. The transmittance of the atmosphere can be predicted by using the scaling factor provided in this figure. The scaling factor corresponds to the equivalent absorber amount, or equivalent path

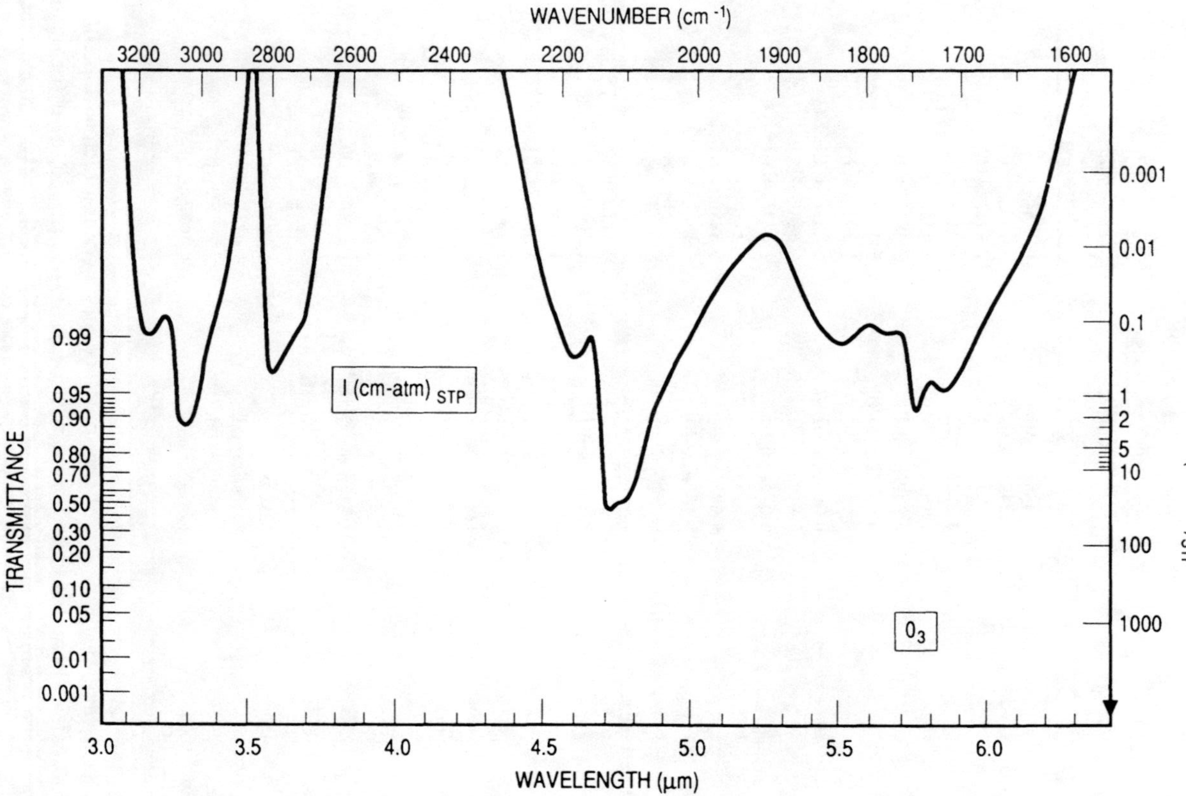

Figure 3.8 Prediction chart for ozone transmission (3–6.4 μm). The STP designation refers to standard temperature and pressure.

length essential in the calculation of the transmittance. Because a constant pressure atmosphere path is easier to handle than a slant atmospheric path (variable pressure path), an "equivalent constant pressure path," defined as the constant pressure path that has the same transmittance at each wavelength as does the actual slant path, is used.

The scaling factor needed in predicting the transmittance is obtained by the equivalent sea-level path lengths and plotted. These are called *scaling charts*. An example of this is given in Figure 3.9 for the ozone layer transmission.

The scaling factors are determined from scaling charts for H_2O, O_3, uniformly mixed gases, nitrogen continuum, H_2O continuum, molecular scattering, and aerosol

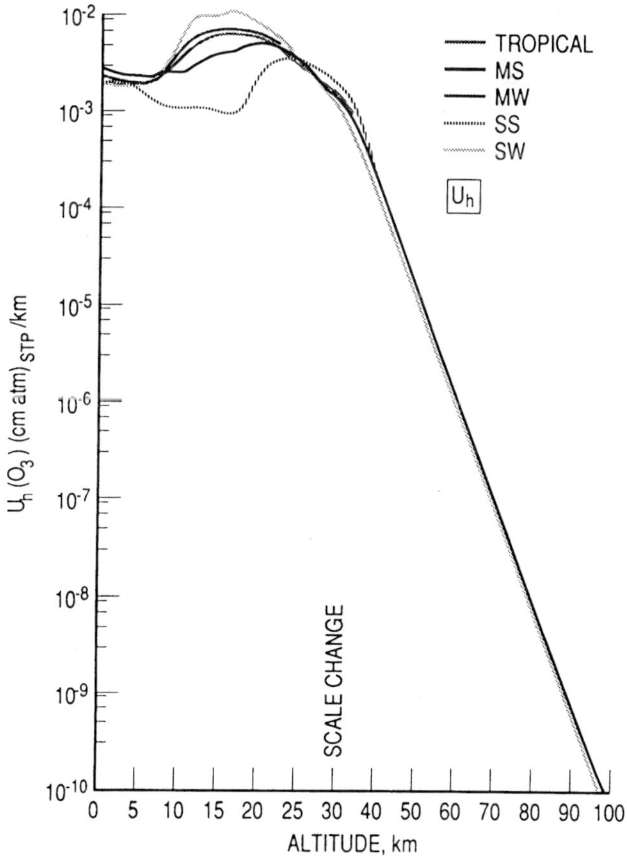

Figure 3.9 Equivalent sea-level path length of ozone as a function of altitude for horizontal atmospheric paths. The STP designation refers to standard temperature and pressure.

extinction, respectively. These values are then used in the prediction charts for the calculation of a transmission coefficient for the specific constituent gas or compound and the specific wavelength.

In cases involving several constituent elements, the total transmittance will be the product of individual values. For example, in the case of H_2O, uniformly mixed gases, and aerosol the total transmittance will be

$$\tau_{total} = \tau_{H_2O} \times \tau_{unif.\ gases} \times \tau_{aerosol} \tag{3.12}$$

The LOWTRAN computer code, widely in use today, started as a digitized version of the prediction charts discussed previously for graphical calculations of atmospheric transmittance. It became more complex as it developed from LOWTRAN 1 to LOWTRAN 6. LOWTRAN was continually modified and improved as new data or models became available. For instance, various aerosol models were added to the original LOWTRAN code, water vapor continuum data was revised, scattering models were added, and so on. More updated LOWTRAN versions also can calculate the atmospheric emittance. These modifications improved the accuracy of the LOWTRAN predictions, but they also added complexity. As the number of aerosol models increased, it became more difficult for the user to choose among the various models for a particular application. Currently, LOWTRAN 7 is under development. Again, many new features are being added, such as multiple scattering of solar and thermal radiation and separate profiles and band model parameters for CO_2, H_2O, O_3, CH_4, and so on.

A comparison of the LOWTRAN calculations of the transmittance with actual measurements is shown in Figure 3.10. Excellent agreement has been achieved, although the LOWTRAN transmittance curve lacks detailed spectral features.

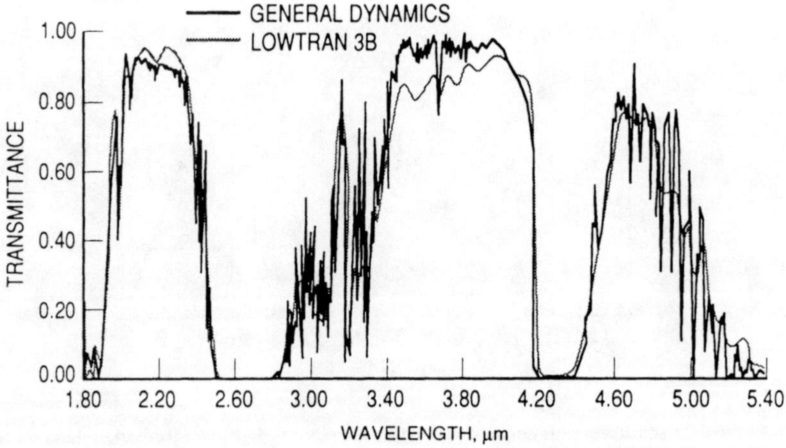

Figure 3.10 Comparison of LOWTRAN calculations with field measurements.

3.8.3 Other Transmission Codes

Other transmission codes have also been reported. These codes are more or less derivative of the HITRAN and LOWTRAN methods.

REALTRAN consists of a series of real-time atmospheric transmission codes, developed at the Naval Research Laboratory, tailored to specific applications [7]. Codes for both laser and broadband applications have been developed based on the HITRAN and LOWTRAN transmission codes.

Another transmission code know as LINETRAN has been developed by Grumman Aerospace Corporation for long-range transmittance calculations [8]. It is a line-by-line code for low-altitude transmission at long ranges. LINETRAN has been developed particularly for the performance predictions of long-range IR sensors.

3.8.4 Quick Atmospheric Transmission Loss Computation

In a report by Obert et al. [9] atmospheric transmittance of thermal imaging systems in the 8 to 12 μm region is given as a function of range and other parameters including visibility range, air temperature, and relative humidity. This table is reproduced here as Figure 3.11.

Selected conditions from this table were converted to decibel units and plotted to obtain transmittance values per kilometer, as shown in Figures 3.12 and 3.13. Figure 3.12 represents transmittance values for clear weather conditions of visibility greater than 4 km, and Figure 3.13 represents conditions of 1 km visibility under fog, with air temperatures of less than and greater than 10°C [10].

From Figures 3.12 and 3.13, it is seen that the atmospheric loss versus range data points represent constant slope lines; that is, atmospheric transmittance in decibels per kilometer remains constant. The transmittance values of Figure 3.12 are summarized in Table 3.6. As seen in this table, the atmospheric transmittance for clear air with visibility greater than 4 km varies between 0.72 and 0.97, depending on relative humidity and air temperature.

From Figure 3.13, the atmospheric attenuation for fog, with 1 km visibility, ranges between 3 and 4 dB/km, depending on the air temperature; that is,

air temperature $>$ 10°C (a factor of 0.40)	4 dB/km
air temperature $<$ 10°C (a factor of 0.50)	3 dB/km

In a similar fashion, atmospheric transmittance in dB/km can be computed for snow and rain conditions; using the values of Figure 3.11 for snow conditions, the atmospheric transmittance will range from 7 to 17 dB/km, depending on visibility and temperature parameters. In rainy conditions, the atmospheric transmittance will

WEATHER OBSERVABLES			TRANSMISSION (%) FOR TARGET RANGE FROM 0.5–15.0 KILOMETERS																	
VR(km)	τ_{AIR} (°C)	RH(%)	0.5	0.75	1.0	1.5	2.0	2.5	3.0	3.5	4.0	4.5	5.0	6.0	7.0	8.0	9.0	10.0	12.5	15.0
FOG																				
0.25	ALL	ALL	0	0	0	0	0	0	0	0	0	0	0	0	0	0	0	0	0	0
0.25 – 0.35	ALL	ALL	5	1	0	0	0	0	0	0	0	0	0	0	0	0	0	0	0	0
0.35 – 0.45	ALL	ALL	16	6	3	0	0	0	0	0	0	0	0	0	0	0	0	0	0	0
0.45 – 0.70	≥10	ALL	38	24	15	6	2	1	0	0	0	0	0	0	0	0	0	0	0	0
0.45 – 0.70	<10	ALL	43	29	19	9	4	2	1	0	0	0	0	0	0	0	0	0	0	0
0.70 – 0.90	≥10	ALL	53	39	29	16	9	5	3	2	1	1	0	0	0	0	0	0	0	0
0.70 – 0.90	<10	ALL	60	47	37	23	14	9	5	3	2	1	1	0	0	0	0	0	0	0
0.90 – 1.25	≥10	ALL	62	49	40	26	17	11	7	5	3	2	1	0	0	0	0	0	0	0
0.90 – 1.25	<10	ALL	70	59	50	36	26	19	13	10	7	5	4	2	1	1	0	0	0	0
1.25 – 1.75	≥10	ALL	72	62	53	40	30	23	17	13	10	8	6	3	2	1	1	0	0	0
1.25 – 1.75	<10	ALL	81	73	67	56	46	39	33	27	23	19	16	12	8	6	4	3	1	1
1.75 – 2.25	≥10	ALL	75	66	59	46	36	29	23	18	15	21	10	6	4	3	2	1	0	0
1.75 – 2.25	<10	ALL	85	79	74	64	56	50	44	39	34	30	27	21	16	13	10	8	5	3
2.25 – 3.00	≥10	ALL	77	68	61	49	40	32	26	21	18	14	12	8	6	4	3	2	1	0
2.25 – 3.00	<10	ALL	87	82	77	69	61	55	50	45	40	36	33	27	22	18	15	12	8	5
3.00 – 4.00	≥10	ALL	78	70	62	51	41	34	28	23	19	16	13	10	6	5	3	2	1	0
3.00 – 4.00	<10	ALL	88	83	79	71	64	58	53	48	44	40	37	31	26	22	18	15	10	7
CLEAR																				
>4.00	>10	80 – 100	82		69	59	50	43	37	32	28	24	21	16	12	9	7	6	3	1
>4.00	>10	50 – 80	87		77	69	63	57	51	46	42	38	35	29	24	20	17	14	9	5
>4.00	>10	<50	96		93	91	89	87	85	84	82	81	79	76	74	72	72	69	63	57
>4.00	<10	>50	94		89	86	82	79	76	73	71	68	66	62	58	54	51	48	40	33
>4.00	<10	<50	98		96	95	93	92	91	90	89	88	87	86	85	83	82	81	77	73
SNOW																				
1.00	≤10	ALL	6		0	0	0	0	0	0	0	0	0	0	0	0	0	0	0	0
1.50	≤10	ALL	15		2	0	0	0	0	0	0	0	0	0	0	0	0	0	0	0
2.00	≤10	ALL	24		6	1	0	0	0	0	0	0	0	0	0	0	0	0	0	0
3.00	≤10	ALL	38		15	6	1	0	0	0	0	0	0	0	0	0	0	0	0	0
4.00	≤10	ALL	47		20	11	5	3	1	0	0	0	0	0	0	0	0	0	0	0
RAIN																				
1.00	ALL	ALL	5		0	0	0	0	0	0	0	0	0	0	0	0	0	0	0	0
1.50	ALL	ALL	17		2	0	0	0	0	0	0	0	0	0	0	0	0	0	0	0
2.00	>10	ALL	25		7	2	0	0	0	0	0	0	0	0	0	0	0	0	0	0
2.00	≤10	ALL	28		8	2	1	0	0	0	0	0	0	0	0	0	0	0	0	0
3.00	>10	ALL	35		13	5	2	1	0	0	0	0	0	0	0	0	0	0	0	0
3.00	≤10	ALL	44		18	8	3	1	0	0	0	0	0	0	0	0	0	0	0	0
4.00	>10	<50	42		26	13	7	3	1	0	0	0	0	0	0	0	0	0	0	0
4.00	≤10	<50	52		32	18	10	6	3	2	1	0	0	0	0	0	0	0	0	0
5.00	>10	<50	54		31	18	10	6	4	2	1	0	0	0	0	0	0	0	0	0
5.00	≤10	<50	62		38	24	15	10	6	4	2	0	0	0	0	0	0	0	0	0

Figure 3.11 Atmospheric transmittance for an 8–12 μm thermal imaging system. (From [9].)

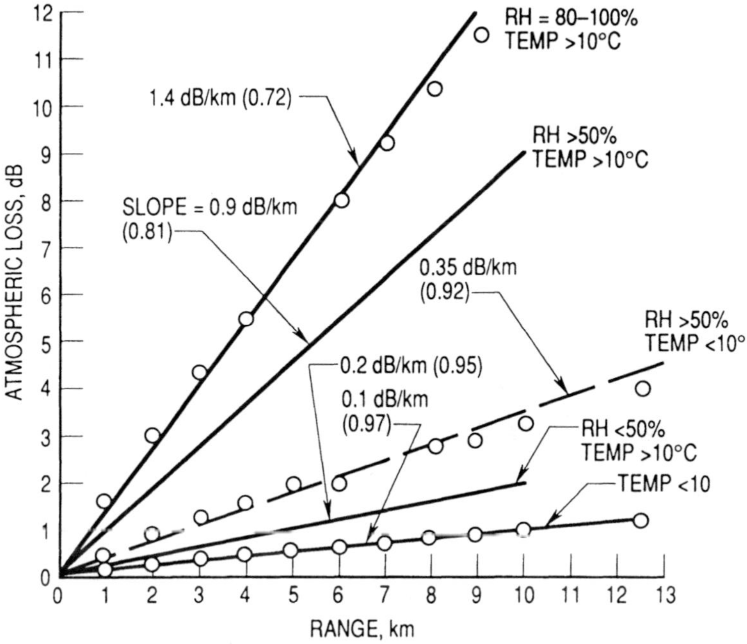

Figure 3.12 Atmospheric attenuation for clear weather with visibility range >4 km.

vary between 4 and 17 dB/km, depending on the values of the same visibility and temperature parameters.

For the purposes of illustration, consider the second line of the clear case of Figure 3.11. In this case, the visibility range is greater than 4 km, the air temperature is greater than 10°, and the relative humidity ranges between 50 and 80%. From Table 3.6, this corresponds to an atmospheric attenuation loss of 0.9 dB/km.

Table 3.6
Atmospheric Transmittance for Clear Weather with Visibility Range Greater than 4 km

Temperature	*>10°C*			*<10°C*	
Relative Humidity	*<50%*	*>50%*	*80–100%*	*<50%*	*>50%*
Atmospheric attenuation Factor/km	0.95 (0.2 dB)	0.81 (0.9 dB)	0.72 (1.4 dB)	0.97 (0.1 dB)	0.92 (0.35 dB)

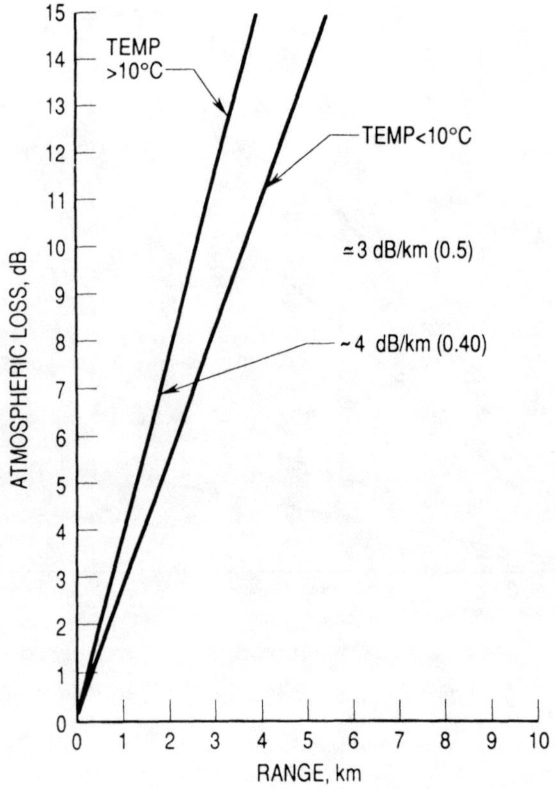

Figure 3.13 Atmospheric attenuation for fog with visibility range ≈1 km.

Example

An example of the atmospheric loss calculation using the preceding method is given in Figure 3.14. This figure shows the signal to noise ratio as a function of detection range for a hypothetical infrared system in the 8 to 12 μm wavelength region. It is seen that the clear path S/N ratio decreases from 50 dB to 10 dB as the detection range increases from 1 to 100 km. Assuming a 0.9 dB/km atmospheric attenuation loss from Table 3.6, the S/N ratio curve with atmospheric attenuation loss of Figure 3.14 can be drawn by subtracting attenuation loss at each range value. Using this figure and assuming a 10 dB threshold setting, the actual detection range of the system, including atmospheric attenuation, will be about 17 km. The unattenuated detection range is 100 km.

Atmospheric attenuation values in the 3 to 5 μm region of the spectrum are close to the values given in Table 3.6. In calculation these values can be used as a reasonable approximation.

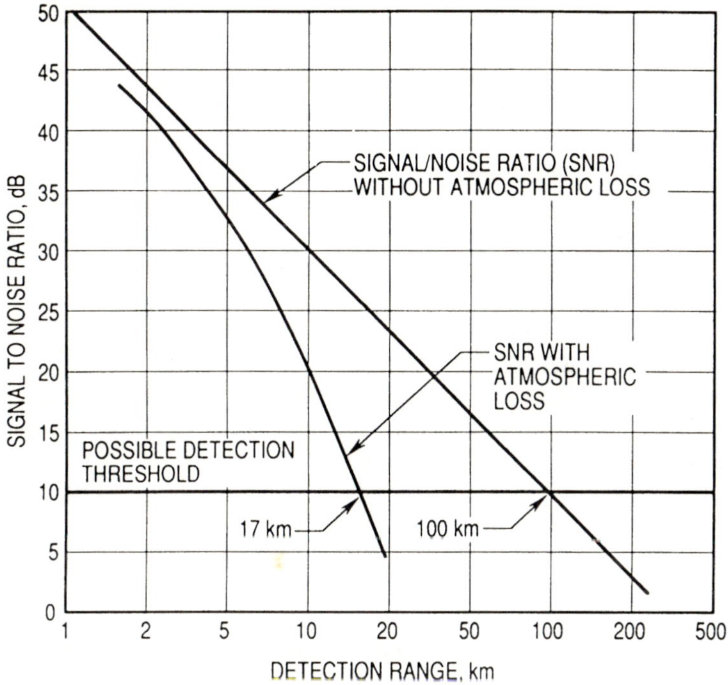

Figure 3.14 Example of quick atmospheric loss computation.

3.9 ATMOSPHERIC TURBULENCE

Atmospheric turbulence is the irregular air movement in which the magnitude and direction of the velocity of air varies randomly at any point; It is the cause of phenomena such as the twinkling of stars at night, the shimmering of distant objects on a hot day, and mirages caused by bending optical rays near a heated surface. Atmospheric turbulence arises when air parcels or turbulent eddies of slightly different temperatures are mixed by wind and convection. It degrades the performance of imaging or laser beam projection systems.

Theoretical investigations on the effects of atmospheric turbulence on optical transmission have been carried on for a number of years. The classic work on wave propagation is by Tatarski [11]; Tatarski's work has since been refined and extended by others to cover the case of wave beam propagation [12–15]. Although there are still problems with the theory, the existing theoretical foundations are adequate to yield reasonable first-order estimates. In addition, a number of experiments have been run to directly quantify the effects and check the theory [16, 17].

3.9.1 The Role of Refractive Index Fluctuations in the Atmosphere

In general, turbulence-induced effects are caused by random refractive index fluctuations [18]. When light propagates through a turbulent atmosphere, the refractive index varies throughout the medium, and the waves of light are affected by this fluctuation. Effects on optical systems include motion, blur, and scintillation in the image. These fluctuations can be identified with turbules or turbulent eddies having characteristic dimensions (or scale sizes), ranging from extremely large, L_0, to extremely small l_0. The letters L_0 and l_0 are called the outer and the inner scale of turbulence, respectively.

The changes in the refractive index of air are due almost entirely to temperature fluctuations. Temperature fluctuations introduced by effects such as the heating of Earth's surface are broken up by the wind and mixed until temperature fluctuations of all scale sizes are developed. During the daytime, solar radiation heats Earth's surface, causing a rise in the air temperature near the ground. The cooler, denser air at higher altitudes descends to replace the warmer air near the surface. This temperature fluctuation is introduced on a large scale by the solar radiation heating of the surface, creating large-scale turbulent eddies. Large eddies break up into smaller eddies, mixing the atmosphere. Through this process, the density of air fluctuates, and therefore the refractive index of air also varies. The energies of larger eddies transfer to smaller eddies, and the turbulence energy dissipates. The dissipation of this energy starts at the inner scale of turbulence, l_0.

Figure 3.15 illustrates the effects of the random fluctuations of the refractive index of air on the propagation of a laser beam for two different cases [19]. Turbules or turbulent eddies having scale sizes greater than the beam diameter act as "weak

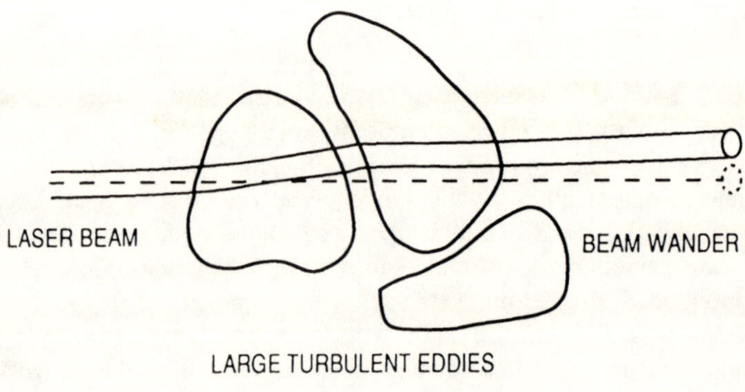

(a)

Figure 3.15 Effects of atmospheric turbulence on laser beam propagation for (a) scale sizes of turbulent eddies > beam diameter and (b) scale sizes of turbulent eddies < beam diameter.

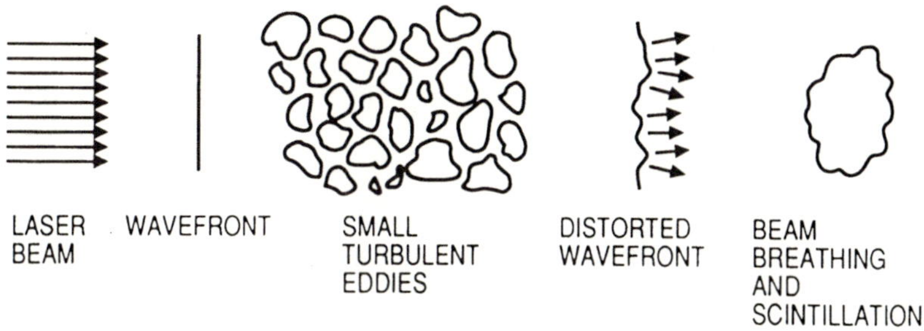

(b)

Figure 3.15 continued.

lenses" that deflect the beam as a whole with no change in beam diameter. These random changes in beam direction are called *beam wander* (see Figure 3.15(a)). Turbulent eddies having scale sizes less than the beam diameter lead to diffractive and refractive effects that cause variations in the beam size known as *beam breathing* and intensity fluctuations known as *scintillation* (see Figure 15(b)). Both cases in Figure 3.15 can occur simultaneously.

3.9.2 Parameters Characterizing Atmospheric Turbulence

The effects of atmospheric turbulence on light propagation are random and complicated. To describe these turbulence effects, statistically averaged quantities are used for a reasonable understanding of the process.

The fluctuation of the refractive index of air can be generally characterized by a refractive-index structure function, $D_n(r)$, given by

$$D_n(r) = \langle (n_1 - n_2)^2 \rangle \tag{3.13}$$

where n_1 and n_2 are the refractive indexes at two points separated by a distance r, and $\langle \, \rangle$ denotes ensemble averaging. Equation (3.13) represents the mean-square increment of the refractive index between the two points. Similarly, the fluctuations in the temperature can be generally characterized by a temperature structure function, $D_T(r)$:

$$D_T(r) = \langle (T_1 - T_2)^2 \rangle \tag{3.14}$$

where T_1 and T_2 are the temperatures at two points separated by distance r.

The structure functions $D_n(r)$ and $D_T(r)$ are described in interval $l_0 \ll r \ll L_0$ by the Kolmogorov-Obukhov two-thirds law, which states that these values are proportional to the distance to the 2/3 power; that is,

$$D_n(r) = C_n^2 \, r^{2/3}$$
$$D_T(r) = C_T^2 \, r^{2/3} \qquad (3.15)$$

where

C_n^2 is the refractive-index structure parameter,
C_T^2 is the temperature structure parameter.

The two-thirds power dependence of $D_n(r)$ and $D_T(r)$ on r holds for the region $l_0 \ll r \ll L_0$, called the *inertia subrange*.

The single most important parameter that characterizes the effects of turbulence on radiative transfer is the atmospheric refractive-index structure parameter, C_n^2, which is related to the temperature structure parameter, C_T^2, by the following expression [19]:

$$C_n^2 = \left[79 \times 10^{-6} \frac{P}{T} \right]^2 C_T^2 \qquad (3.16)$$

where

P is the atmospheric pressure in millibars,
T is the temperature in K.

The value of the atmospheric structure coefficients, C_n^2, varies as a function of altitude above ground. This value is generally maximum at ground level and decreases with altitude. Hufnagel and Stanley [20] have given the following expression for C_n^2:

$$C_n^2 = 4.2 \times 10^{-14} \, h^{-1/3} \exp(-h/h_0) \qquad (3.17)$$

where

h is the altitude above ground in meters,
h_0 is 3,200 m.

For most ground applications, the value of C_n^2 can generally be assumed to be constant. However, for air-to-ground applications, the variations may be significant enough to use eq. (3.17).

The parameter C_n^2 is also a function of the time of day. The heating of Earth's surface during the day creates temperature fluctuations, and hence higher values of C_n^2 are calculated. Therefore, the strength of turbulence varies over the diurnal cycle with greater turbulence (or C_n^2) typically occurring during the day, as shown in Figure 3.16.

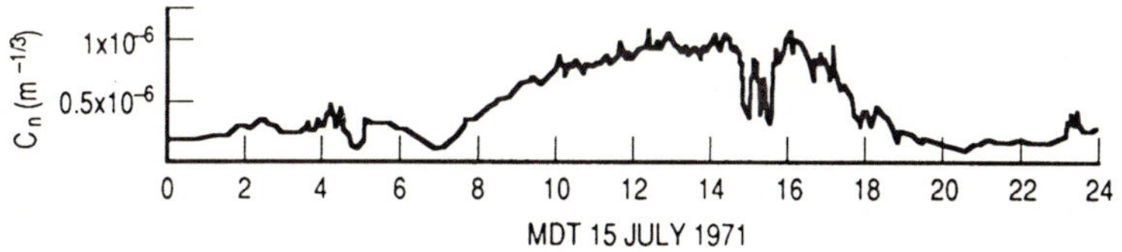

Figure 3.16 Plot of a typical diurnal variation of the refractive index structure parameter near the ground on a clear, sunny day (the drops in C_n at hour 15 are due to clouds).

Typical values of C_n^2 range from about 10^{-17} $m^{-2/3}$ for weak turbulence to about 10^{-13} $m^{-2/3}$ for strong turbulence.

The scale sizes are related to the power spectrum of refractive index fluctuations through a turbulence model. The models most often used are Tatarski's,

$$\phi_n(k) = 0.033\, C_n^2\, k^{-11/3} \exp(-k^2/k_m^2) \tag{3.18}$$

and the modified or Van Karmen spectrum,

$$\phi_n(k) = \frac{0.033\, C_n^2 \exp(-k^2/K_m^2)}{[k^2 + (L_0/2\pi)^{-2}]^{11/6}} \tag{3.19}$$

where

$\phi_n(k)$	is the three-dimensional power spectrum of the refractive index fluctuations,
k	is $2\pi/\lambda$,
L	is the scale (or dimension) of turbulence,
C_n^2	is the atmospheric structure coefficient or "constant,"
K_m	is $5.92/l_o$,
l_0	is called the *inner scale* of turbulence,
L_0	is called the *outer scale* of turbulence.

Figure 3.17 shows a plot of $\phi(k)$ versus k. Typical values for L_0 vary from the height above ground to 100 m or more for the upper atmosphere. The inner scale length,

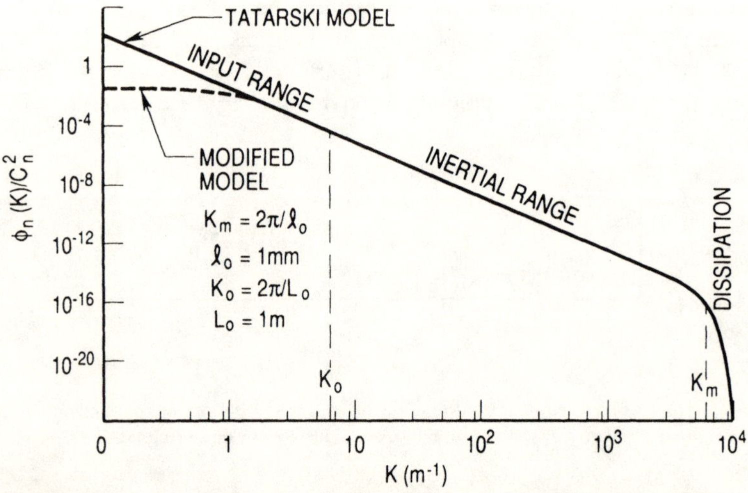

Figure 3.17 Three-dimensional spectrum of the refractive index fluctuations.

l_0, defines the turbulence scale sizes for which viscous dissipation converts the energy of turbulence into heat, and it is typically 1 mm near the ground to a few millimeters at altitude.

3.9.3 Effects of Atmospheric Turbulence on the Performance of Optical Imaging Systems

Atmospheric turbulence has adverse effects on the propagation of light and can cause a serious degradation in the performance of optical imaging systems. Like the diffractive effects of the optics, which smear the image-plane irradiance of a point source object, the presence of turbulence in the atmosphere also degrades the system performance. The turbulent atmosphere, as discussed previously, has both refractive and diffractive effects, such as image blur and image wander, caused by the inhomogeneities of the index of refraction [14].

Therefore, two sources of image degradation, when light propagates from the object plane to the image plane, are

1. Diffraction caused by the optics,
2. Refraction and diffraction caused by atmospheric turbulence.

It is useful to have a model that predicts the performance of optical imaging systems in the presence of atmospheric turbulence. The *optical transfer function* (OTF) can be used as a fundamental quantity that describes the effects of atmospheric turbulence on the resolution of an optical imaging system or on the spot size of a laser beam projection system. Fried [21] has derived an expression for the OTF of the atmosphere, which is dependent on the strength of atmospheric turbulence. Others, after using this theoretical model and comparing it to experimental results, have concluded that a good agreement exists between theory and experiment.

The model by Fried for spherical waves is given by the expression [21]

$$\mathrm{OTF} = \mathrm{e}^{-\{21.62 C_n^2 R \lambda^{-1/3} f_s^{5/3} [1-\alpha(f_s \lambda / D_0)^{1/3}]\}} \tag{3.20}$$

where

C_n^2 is the refractive index structure parameter,
R is the range or path length,
λ is the wavelength,
f_s is the angular spatial frequency in cycles/rad,
D_0 is the diameter of aperture,

$$\alpha \begin{cases} 0 \text{ for a long-term OTF} \\ 1/2 \text{ for a far-field, short-term OTF} \\ 1 \text{ for a near-field, short-term OTF} \end{cases}$$

Equation (3.20) gives the OTF expression of the atmosphere limited in general to two cases: short term and long term.

The short-term OTF is a short exposure OTF, in which the image is obtained over a short time interval. If a series of short-term OTFs are taken from a point source, there would be beam spreading and beam wandering effects at the image plane caused by atmospheric turbulence. The average short-term OTF is obtained by shifting the center of the area of each *point spread function* (PSF) to a common point and then averaging and taking the Fourier transforms of the successive PSFs. The shifts required to center the image are known as *image wander* to *beam wander*. This averaging procedure has the effect of centering the image and removing the beam wander, and the short-term OTF is often called the *beam spread*.

The long-term OTF is obtained from a long exposure time, so that every turbulence configuration has been taken into account on a single exposure. Average long-term OTF is therefore the same as that of a single long exposure, and there is no need for the shifting process discussed previously. Long-term OTF, therefore, refers to both the *beam spread* (short-term OTF) and *beam wander*.

The image wander, expressed in radians, is also useful in describing the performance of an imaging system. Image wander can be obtained by calculating the *root mean square* (rms) deviation from the mean of the magnitudes of the shifts needed to center the PSFs in the average short-term OTF calculation. The variance of image wander is simply the square of rms image wander, and it is given by Fried's model for spherical waves as [21]

$$\langle\theta^2\rangle = 1.12\, C_n^2\, R\, D_0^{-1/3} \tag{3.21}$$

where C_n^2, R, and D_0 are defined in eq. (3.20).

Whether the condition is near field or far field in eq. (3.20) can be determined by the expression

$$\Omega = \frac{k\, D_0^2}{R} \tag{3.22}$$

where

- Ω is the Fresnel number,
- k is the wavenumber ($k = 2\pi/\lambda$),
- D_0 is the aperture diameter,
- R is the path length.

In eq. (3.22), $\Omega \gg 1$ corresponds to the near-field condition, and $\Omega \ll 1$ corresponds to the far-field condition.

To test the theoretical model by Fried, the OTF of the atmosphere can be measured experimentally. Crittenden et al. [22], for example, used slit scanning optics to scan the image of a distant point source (a laser source). The result is a one-dimensional spread function due to diffraction effects and aberration of the optics as well as atmospheric effects. The contributions of the one-dimensional spread function from the optics and instruments are removed to give the OTF of the atmosphere. A two-dimensional point spread function is obtained from the one-dimensional PSF assuming a circularly symmetric PSF. Some results of this work are shown in Figure 3.18. The square dots are the data points from the atmospheric OTF measurements and the solid lines are the best fit curves of the data points using the Fried model, eq. (3.20), with the adjustable parameter being C_n for best fit. The curve fit is quite good, especially at the lower frequency scale. In Figure 3.18, the values of C_n inferred by the short-term OTF and the long-term OTF are $9.16 \times 10^{-8}\ \text{m}^{-2/3}$ and $9.95 \times 10^{-8}\ \text{m}^{-2/3}$, respectively, for the following pertinent parameters:

$$\lambda = 0.488\ \mu\text{m}$$
$$R = 18.8\ \text{km}$$
$$D_0 = 0.4572\ \text{m}$$

A third value of C_n is obtained from the image wander, see eq. (3.21). This value is $11.0 \times 10^{-8}\ \text{m}^{-2/3}$ in Figure 3.18. The validity of the Fried theory, then relies

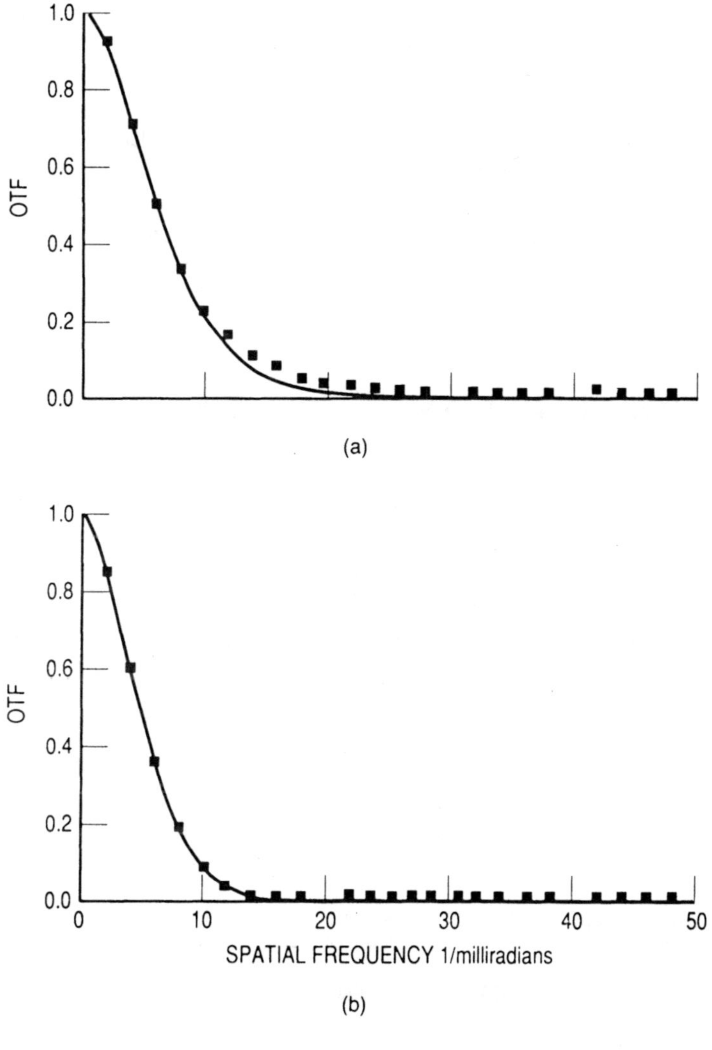

Figure 3.18 Comparison of thermal measurements of short- and long-term optical transfer function: (a) short-term OTF for 0.488 μm; (b) long-term OTF for 0.488 μm.

on the consistencies of these values of C_n. The example in Figure 3.18 shows three values of C_n that are relatively consistent with each other. Other experimental data with different parameters show the same type of consistency between the values of C_n, proving the validity of the Fried theory. The Fried model for the atmospheric OTF, then, is a useful tool for the description of the turbulence induced effects on optical imaging systems.

Comparison of optical transfer function of Figure 3.18 of this chapter with atmospheric turbulence with similar optical response curves given in Chapter 4 (Figure 4.12) will show that turbulence, as expected, degrades system performance. Under normal nonturbulent conditions the optical transfer function starts at unity at the origin, as given in Figure 3.18, and degrades considerably slower as one proceeds to higher values of spatial frequency. Note that spatial frequency is related to the diffraction limit of the optical lens with higher values of spatial frequency representing "off-boresight" values, with the extreme values being the "null" value of the diffraction limit. Using this reasoning another way of looking at the turbulence effect may be to consider it a loss of signal-to-noise ratio similar to atmospheric effects considered previously.

3.10 ADAPTIVE OPTICS AND SIGNAL PROCESSING

Adaptive methods, whereby the system responds to dynamic changes of signal and background, have been applied both in design of optical systems and in signal processing methods. Adaptive optics may include corrections dictated by changes in background and/or target signal so to optimize the behavior of the optical design. Adaptive signal processing methods may include variable threshold settings as a function of background noise (clutter) to obtain constant false alarm rate. Adaptive methods are usually tailor-made for specific designs under variable operating conditions [26].

REFERENCES

[1] McCartney, E.J., *Optics of the Atmosphere*, New York: John Wiley and Sons, 1976.

[2] Jaimeson, J.A., et al., *Infrared Physics and Engineering*, New York: McGraw-Hill, 1963.

[3] Valley, S.L. (Ed.), *Handbook of Geophysics and Space Environment*, New York: McGraw-Hill, 1965.

[4] McClatchey, R.A., "Atmospheric Transmission Models and Measurements," *Proc. SPIE*, 1979.

[5] Clough, S.A., et al., "Atmospheric Spectral Transmittance and Radiance: FASCOD1B," *Proc. SPIE 277*, 1981.

[6] McClatchey, R.A., "Optical Properties of the Atmosphere," AFCRL-72-0497, 1972.

[7] Haught, K.M., "REALTRAN Real Time Implementation of Atmospheric Codes," *Proc. SPIE 277*, 1981.

[8] Smith, L., and L.T. Hilgeman, "High Resolution Lower Atmospheric Transmission Prediction over Long Paths," *Proc. SPIE 277*, 1981.

[9] Obert, L.P., et al., "Visionics EO Sensor Performance," Arlington, VA: DCS Corporation, June 1981.

[10] Hovanessian, S.A., *Introduction to Sensor Systems*, Norwood, MA: Artech House, 1988, p. 197.

[11] Tatarski, V.I., *Wave Propagation in a Turbulent Medium*, New York: McGraw-Hill, 1960.

[12] Strohbehn, J.W., "Line-of-Sight Wave Propagation Through the Turbulent Atmosphere," *Proc. IEEE*, Vol. 56, August 1968, pp. 1301–1318.

[13] Fried, D.L., and J.B. Seidman, "Laser Beam Scintillation in the Atmosphere," *J. Opt. Soc. Amer.*, Vol. 57, 1967, pp. 181–185.

[14] Nevers, G.E., M.P. Kuster, Jr., and D.L. Fried, "Saturation of Scintillation," *J. Opt. Soc. Amer.*, Vol. 59, No. 4, 1969, pp. 491–492.

[15] Dooling, J.A., and P.M. Livingston, "Behavior of Focussed Beams in Atmospheric Turbulence: Measurements and Comments on the Theory," *J. Opt. Soc. of Amer.*, Vol. 63, No. 7, July 1973.

[16] Kerr, J.R., et al., "Atmospheric Optical Communications Systems," *Proc. IEEE*, Vol. 50, No. 10, October 1970.

[17] Kerr, J.R., and J.R. Dunphy, "Experimental Effects on Finite Transmitter Apertures on Scintillations," *J. Opt. Soc. of Amer.*, Vol. 63, No. 1, January 1973.

[18] Lawrence, R.S., "A Review of the Optical Effects of the Clear Turbulent Atmosphere," *Proc. SPIE 75*, 1976, pp. 2–8.

[19] Weichel, H., "Atmospheric Propagation of Laser Beams," *Proc. SPIE 547*, 1985, pp. 1–15.

[20] Hufnagel, R.E., and N.R. Stanley, "Modulation Transfer Function Associated with Image Transmission Through Turbulent Media," *J. Opt. Soc. Amer.*, No. 54, 1964.

[21] Fried, D.L., "Optical Radiation Through a Randomly Inhomogeneous Medium for Very Long and Very Short Exposures," *J. Opt. Soc. Amer.*, Vol. 56, 1966, p. 1372.

[22] Crittenden, E.C., et al., "Effects of Turbulence on Imaging Through the Atmosphere," *Proc. SPIE 142*, 1978, pp. 130–134.

[23] Zuyev, V.Y., *Propagation of Visible and Infrared Waves in the Atmosphere*, Washington, DC: NASA TTF-707, 1972.

[24] McCartney, E.J., *Absorption and Emission by Atmospheric Gases*, New York: John Wiley and Sons, 1983.

[25] McClatchey, R.A., and A.P. D'Agati, "Atmospheric Transmission of Laser Radiation: Computer Code LASER," AFGL-TR-78-0029, 1978.

[26] Tyson, R.K., Principles of Adaptive Optics, New York: Academic Press, 1991.

Chapter 4
Spectral, Spatial, and Temporal Variations in Infrared Backgrounds

4.1 INTRODUCTION

Perhaps the most difficult problem in the infrared field is the derivation of a credible background radiance model by which to predict the performance of infrared sensors under practical conditions. Unfortunately, to date, the work in generating this model has been spotty, at best. The difficulty in generating a credible background model is due to the large number of parameters, each of which contributes to some variation in this model. These parameters include the spectral band, spatial background structure, geographic location, isolation, viewing and scattering angles, aspect angle, emissivity and reflectivity variation, atmospheric absorption, range, temporal variation, including background motion, the effect of day and night, as well as seasonal variation, and finally, the effect of sensor parameters, including optics-detector spectral and spatial response and detector-electronic temporal response.

The effects of some of these parameters, such as optics, detector, and electronics, such as the spatial variation of background radiance in different spectral bands or the effect of atmospheric absorption and range, cannot be precisely predicted. Therefore, the popular approach has been to group all of these parameters together and construct a one or two-dimensional random function for the background radiance. The generation of such a model is, in most cases, difficult because the statistics of the background radiance function do not conform to the usually well-behaved stationary and ergodic random processes. The fact that background radiance, as observed by a scanning sensor, may produce a function whose statistics could be different from one spatial location to another makes this function nonstationary in space. Furthermore, these statistics change with time, either because of background motion or variation in atmospheric constituents with time, such as water, vapor, haze, or fog, indicating that background radiance functions are stationary neither in time nor in space.

The irony of the formulation of a background model is in the fact that the complex formulation of the background spatial, spectral, and temporal radiance function for the general case makes application of this model impractical. An oversimplification of this formulation, however, may result in an unrealistic description of this complex function with consequent inaccurate results.

The spectral correlation technique of the spatial radiance function, has been one of the most ill-applied discrimination techniques of IR systems. Sometimes the background spatial function, which shows relatively good correlation in two separate spectral bands, has shown very little or no correlation after processing. This is due to the effects of many parameters that affect the background radiance and has resulted in the failure of spectral correlation techniques in many discrimination applications. However, if the effect of these parameters are known *a priori*, a processing technique may be devised so that spectral correlation can, in some cases, be effectively applied.

The historical approach in generating the background model, unfortunately, has been the slingshot approach. Periodically, a newcomer to the field, possibly with some funding capability, gets interested in this problem and sponsors a particular group to investigate a part of this problem. The results of such an investigation has been mostly a rehash of the work done by previous investigators, resulting in no significant improvements. In most cases, the new experimenter who becomes knowledgeable in the field and discovers the difficulty of the problem is either transferred to another position or decides not to pursue such a difficult task.

It must be recognized that the failure in generating a credible background radiance model as well as reaching an understanding of its application to the design and performance evaluation of IR sensors can result in a great penalty. The history of IR systems performance evaluation in actual field testing proves this claim. Many IR systems developed for a specific function failed in the field tests because the system designer either did not incorporate or did not have the proper background radiance data. It must also be recognized that this area of IR technology is of sufficient complexity that only personnel with extensive IR background, with an understanding of the mathematical formulation, and a sincere devotion to this field should attack the problem.

The purpose of this chapter is to derive a general equation of the background radiance function as explicitly as possible, evaluate the effects of various parameters, and discuss the validity of simplifications of the radiance function whenever this "simplification" is justifiable. The chapter further discusses the important factors in the design and experimentation of background in optical detection systems.

4.2 BACKGROUND SPECTRAL RADIANCE

During the last two decades, several background measurement experiments have been conducted both in the wide and narrow band regions of the spectrum. Wideband measurements refer to specific bands where detectors are operating. These bands, for example, may be in the 3 to 5 μm or 8 to 12 μm region. Narrowband measurements, on the other hand, are normally made with radiometers with a fine spectral resolution. In these cases either a scanning or a nonscanning radiometer may be used. A scanning radiometer transforms the spatial background function to a temporal function at the output of the detector. Therefore, the design of postdetector electronics, including the frequency response, sampling rate, dynamic range, and in particular the calibration of the instrument, becomes critical. The background measurement data so obtained is essential in the design of high-sensitivity and high-performance IR systems, such as search, track, and seeker systems.

A large number of field experiments over the last two decades have been conducted on the spectral radiance of various types of backgrounds. The existence of high-altitude aircraft, satellites, and balloons has facilitated the spectral measurement of various backgrounds.

Figure 4.1 shows the radiation of the sky as seen from Earth [1, 2]. This background radiation in the infrared region is caused by scattering of the sun's radiation

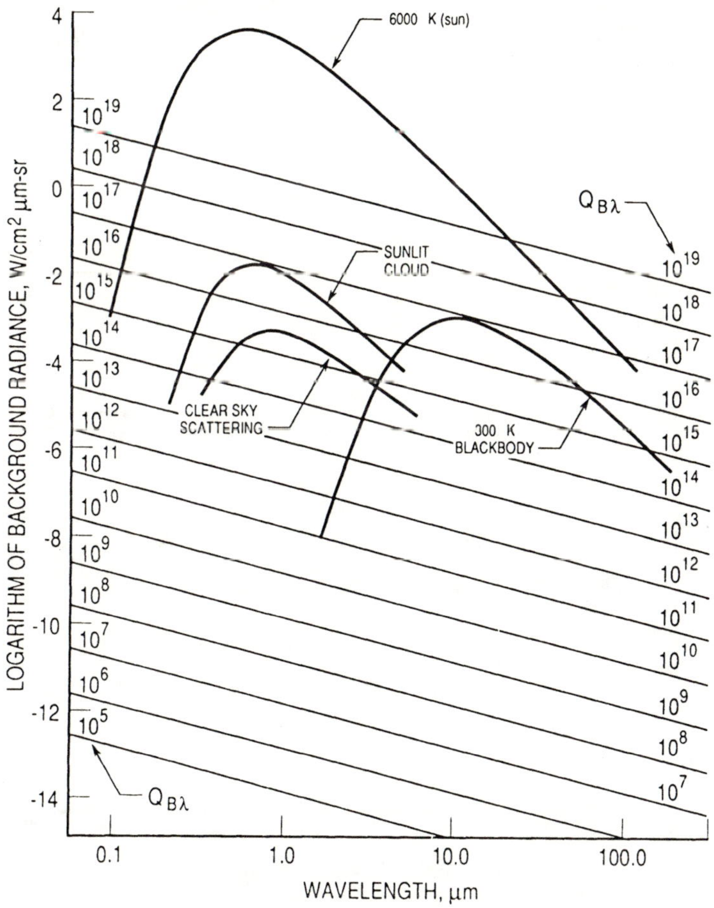

Figure 4.1 Contributions from scattering and atmospheric emission to background radiation.

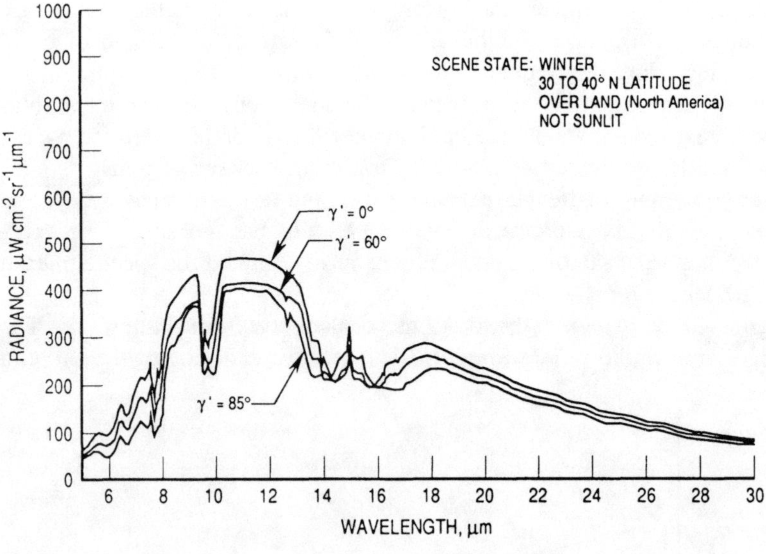

Figure 4.2 The expected value of spectral radiance for three γ' angles.

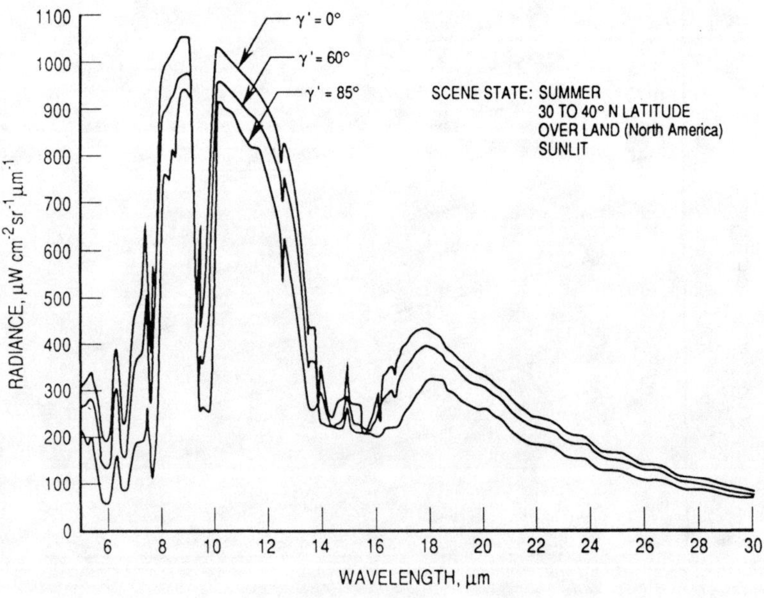

Figure 4.3 The expected value of spectral radiance for three γ' angles.

and emission from atmospheric particles. Note that the logarithm of the background radiance is plotted against wavelength. The units of background radiance are in watts per (cm^2-μm-sr). The figure also contains photon radiance lines $Q_{B\lambda}$ with units of photons per (cm^2-μm-sr-s). These values are obtained by multiplying background radiance by $h\nu$, where h is Planck's constant and ν is the frequency.

Figures 4.2 and 4.3 show the spectral radiance of sunlit ground during winter and summer, respectively [3]. The angle γ' of these figures is zero when the observer is directly overhead and is 90° when looking at the horizon (see Figure 4.4). The spectral band of Figures 4.2 and 4.3 is between 5 to 30 μm. Note also the notch between 9 and 10 μm due to ozone absorption.

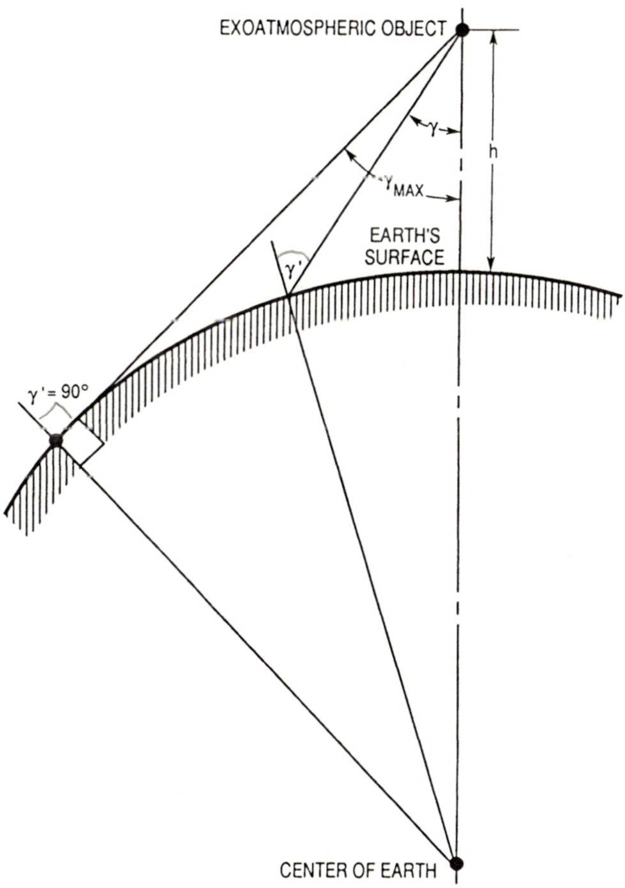

Figure 4.4 Cross-section view of the geometry pertaining to the calculation of the thermal radiance field at an exoatmospheric object.

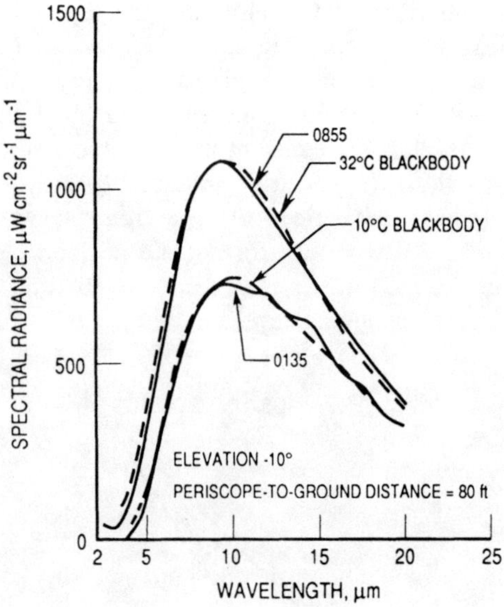

Figure 4.5 Day and night radiances of a grass-covered field.

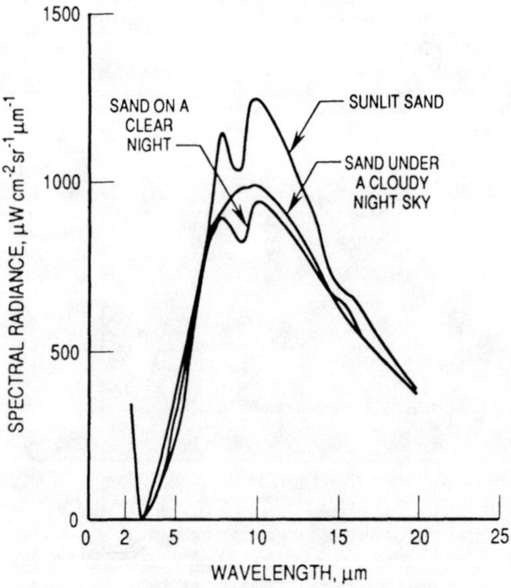

Figure 4.6 Spectral radiance of dry sand (Cocoa Beach, Florida), under three conditions.

Figures 4.5 and 4.6 give the radiance of grass-covered fields and dry sand under a variety of specified conditions [4]. Figure 4.5 gives the radiance of a grass-covered field in two different times of morning and afternoon, and Figure 4.6 gives the radiance of sand for three different conditions: sunlight, cloud, and night.

Thermal emission and spectral radiation for lower spectral wavelength regions are also measured as presented in Figure 4.7. Note the units of this figure are expressed in logarithms of the values. The ground, cloud, and atmospheric emissions peak at a value of 10^{-3} W/(cm^2-sr-μm), whereas the solar reflection peaks at a value of 10^{-2} W/(cm^2-sr-μm) at a wavelength of 0.5 μm as shown in the figure. The reflection values are given for reflectivity values of ρ as specified.

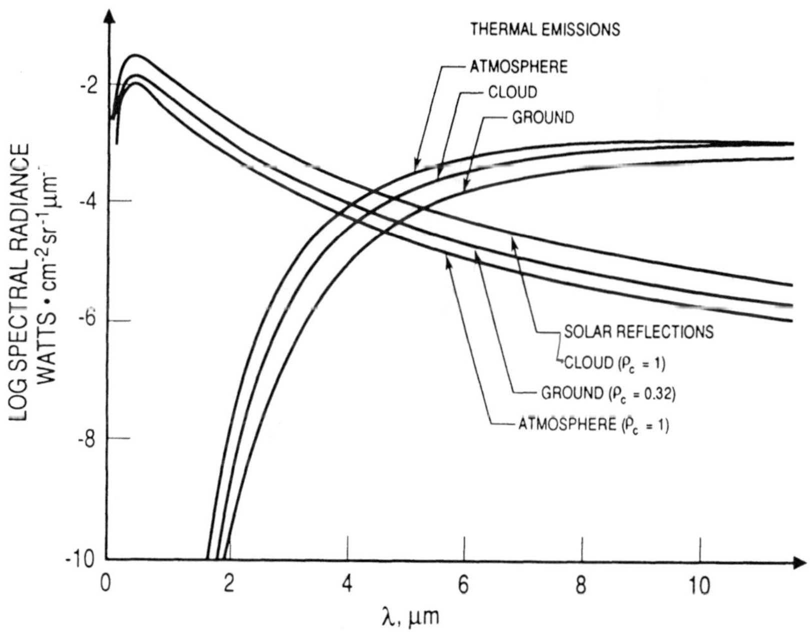

Figure 4.7 Thermal emission and solar reflection at wavelengths below 10 μm (no atmospheric attenuation).

4.3 REFLECTION AND EMISSIVITY CHARACTERISTICS FOR IDENTIFICATION

The reflectance and emissivity characteristics discussed previously were primarily used for system and background noise level calculations. These values are then used to calculate signal-to-noise (background) level and determine subsequent sensor performance. The reflectance and emissivity characteristics can also be used to identify terrain, vegetation, ice cover, oil slicks and so forth.

For example, Figure 4.8 gives the emissivity characteristics of common rocks as a function of wavelength in the region of 8 to 14 μm with limestone having the highest emissivity characteristics.

Figure 4.9 gives visible and near infrared signatures of minerals. The number of discrete spectral bands needed for each mineral identification depends on the location, depths, and shapes of absorption features to be measured.

Similar identification methods may be used in the visible spectral region of blue, green, and red as well as near and far IR spectral regions.

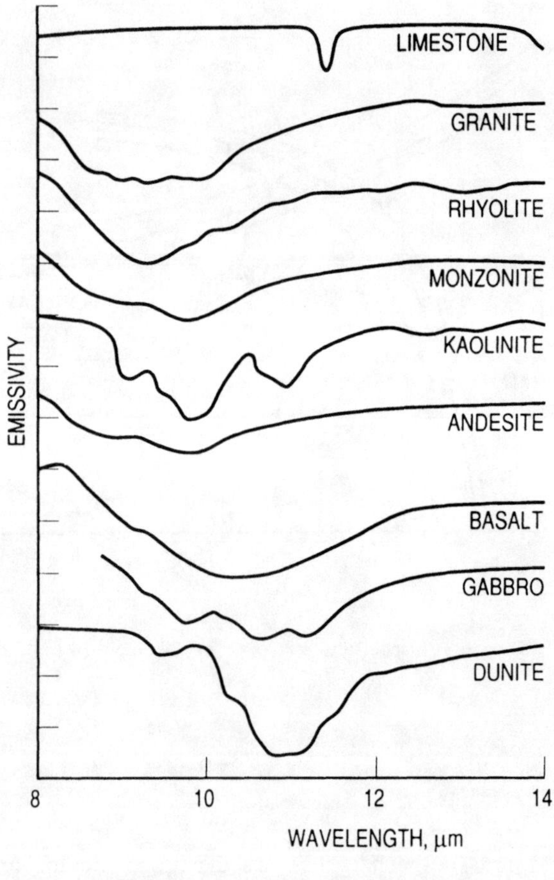

Figure 4.8 Thermal-infrared spectra of common rocks; these major rock types are readily distinguished using the thermal-IR spectral region.

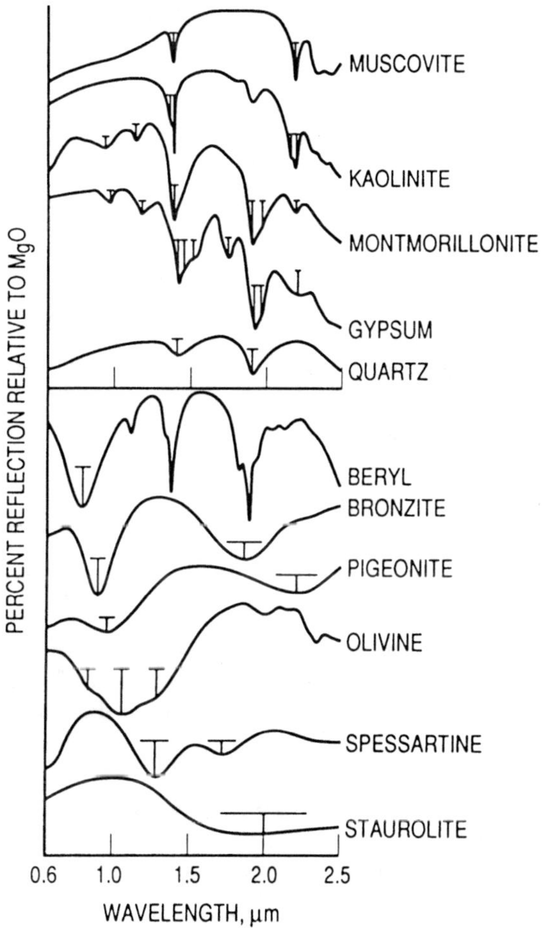

Figure 4.9 Visible and near-infrared spectra of minerals; these spectra provide a unique means of differentiating between many Fe and OH bearing minerals.

Imaging spectrometers are planned for spaceborne and airborne systems with as much as 200 spectral bands in the spectrum region of 0.4 to 2.5 μm for identification of mapped areas, as described in [5].

4.4 BACKGROUND SPATIAL RADIANCE

A general formulation of the spatial background radiance function follows. The integrals of these equations are usually evaluated numerically by using a combination of measured values and mathematical analysis. In studying these equations it is suggested that the reader concentrate on underlying concept rather than mathematics involved.

The background radiance from the three common background types—terrain, sea, and sky (including clouds)—can be written in general as follows:

$$N_t(R, \eta, \varepsilon, \lambda, t) = N_b(R, \eta, \varepsilon, \lambda, t) \cdot \tau(R, \eta, \varepsilon, \lambda, t) + N_{\text{atm}}(R, \eta, \varepsilon, \lambda, t) \quad (4.1)$$

where, from Figure 4.10,

- R is the sensor-target range,
- λ is the wavelength,
- t is the observation time,
- η is the azimiuth angle,
- ε is the elevation angle,
- N_t is the total spectral background,
- N_b is the spectral radiance at the source,
- τ is the spectral atmospheric transmission characteristics in the sensor-background path,
- N_{atm} is the radiance of the atmosphere in this path.

The dependence of the background irradiance at the sensor entrance aperture for range R is mainly due to the atmospheric transmission. The dependence of the background irradiance on the time parameter t is due to the actual change of background radiance with time. The time effects may be due to the motion of the background or the temperature change of the background over time when self-emission of the background is considered. Other temporal effects include atmospheric phenomena, such as evolving fog or atmospheric scintillation.

The spectral incident power at the optical lens will be

$$\frac{\pi D^2}{4R^2} N_t(R, \eta, \varepsilon, \lambda, t) \quad (4.2)$$

The power incident on the detector at a focal length of f, with subtended angles of ψ and ξ can be computed from a consideration of target area subtended by these angles. The subtended target area at range R will be

$$(\psi)\,(\xi)\,R^2 \quad (4.3)$$

where ψ and ξ are detector instantaneous field of view values. Using target area eq. (4.3) in eq. (4.2) we have

$$\frac{\pi D^2}{4}\,(\psi)\,(\xi)\,K_0\,(\lambda)\,N_t\,(R, \eta, \varepsilon, \lambda, t) \quad (4.4)$$

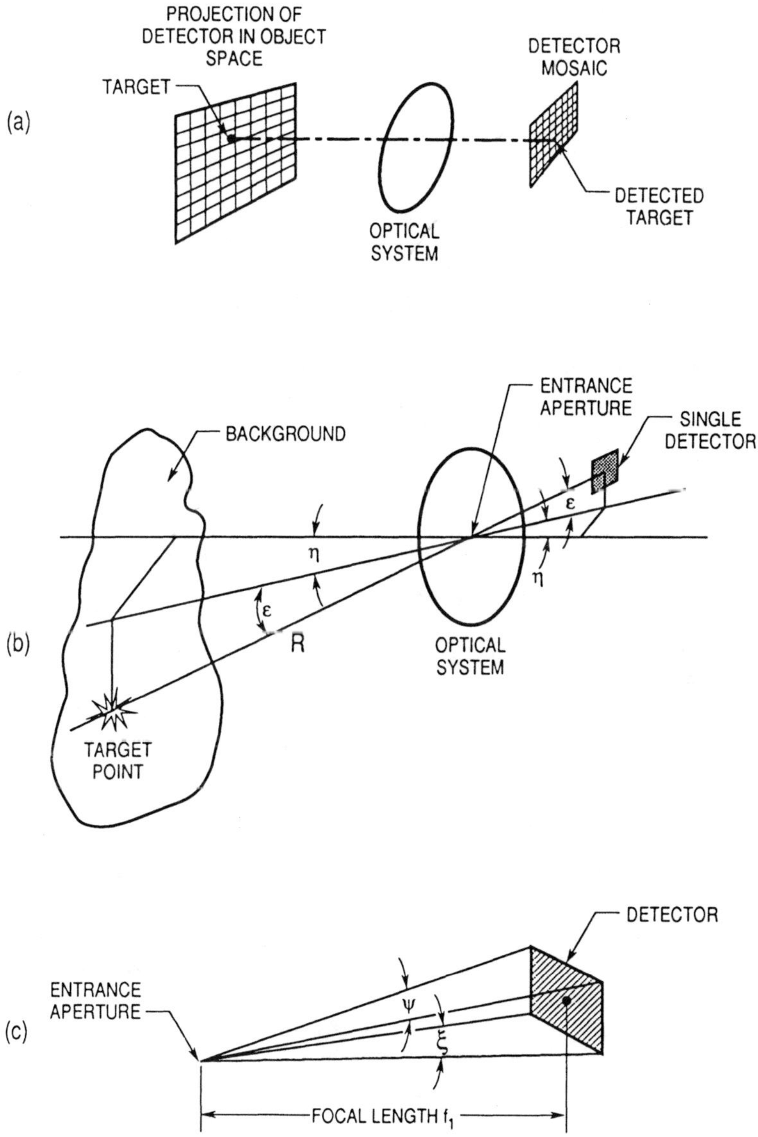

Figure 4.10 (a) Projection of a scene into the focal plane array or detector mosaic; (b) projection of the target into a detector; and (c) geometry of a single detector with respect to entrance.

where $K_0(\lambda)$ is the spectral transmission of the optics. In a number of cases the detector IFOV is expressed in steradian units as the product of IFOV values; that is,

$$\alpha_d = \psi\,\xi \tag{4.5}$$

Using this in eq. (4.4), we get the spectral power at the detector as

$$h_b = \frac{\pi D^2}{4}\,\alpha_d\,K_0(\lambda)\,N_t\,(\rho,\,\eta,\,\tau,\,\lambda,\,t) \tag{4.6}$$

where h_b is the geometrical spectral irradiance of the background at the focal plane, that is, if we assume a (δ) function response for the telescope. For a practical telescope, diffraction and aberration spread this image. Note also the absence of range R in eq. (4.6). The reason is that the subtended field of view angle α determines the area of the background. At longer distances, larger areas are covered; this larger area coverage compensates for R^2 transmission loss, eliminating the dependence of background signal on range.

If the point spread function of the telescope is $g(\eta,\,\varepsilon,\,\lambda)$, where η and ε are angular coordinates of the focal plane as they are observed from telescope entrance aperture of Figure 4.10, the image plane irradiance can be written as

$$h_i(R,\,\eta,\,\varepsilon,\,\lambda,\,t) = h_b(R,\,\eta,\,\varepsilon,\,\lambda,\,t) * g(\eta,\,\varepsilon,\,\lambda) \tag{4.7}$$

where $*$ designates a convolution operation.

Summarizing the results, the total power at the detector at the location η and ε will be

$$P_d(\eta,\,\varepsilon,\,\lambda) = \int_{\varepsilon-\xi/2}^{\varepsilon+\xi/2}\int_{\eta-\psi/2}^{\eta+\psi/2}\int_{\lambda-\Delta\lambda/2}^{\lambda+\Delta\lambda/2} h_i(R,\,\eta,\,\varepsilon,\,\lambda,\,t)\,\mathrm{d}\lambda\,\mathrm{d}\eta\,\mathrm{d}\varepsilon \tag{4.8}$$

The preceding integral will give the total power incident on the detector P_d as a function of time. The detector location being η and ε with respect to the lens axis.

In the background radiance measurements using spectrometers or radiometers, the detector incident power is measured. Then, by using the parameters of the optics, the problem is worked out in reverse and the background radiation value at the source N_b is estimated. This is similar to calibrated instrumentation where the instruments are calibrated against known signals. Following this, the instruments are exposed to the background where the radiance is to be measured. Comparison of the known signal and the measured signal will give a good estimate of background radiance.

Spatial variation of background radiance as observed by a scanning sensor can be caused by two different phenomena:

1. The scan line passes through several different types of background. The scan line may cover, for example, a clear sky then clouds, mountains, water, and so on. In each transition step some abrupt changes in the background radiance may be detected.
2. The scan line passes through only one type of background, terrain background, for example. The spatial variation in this case can be caused by variation in illumination angle, viewing angle, range, and so forth.

The changes in the measured radiance in the first case is actually caused by the differences in radiances of each background. A typical background radiance of two types of backgrounds features (blue sky and water) and their differences are shown in Figure 4.11.

The spatial radiance function as measured in the second case, that is, the spatial variation in one type of background, however, is of vital importance. It is from measurement of only one type of background that the generation of a statistical model can be meaningful.

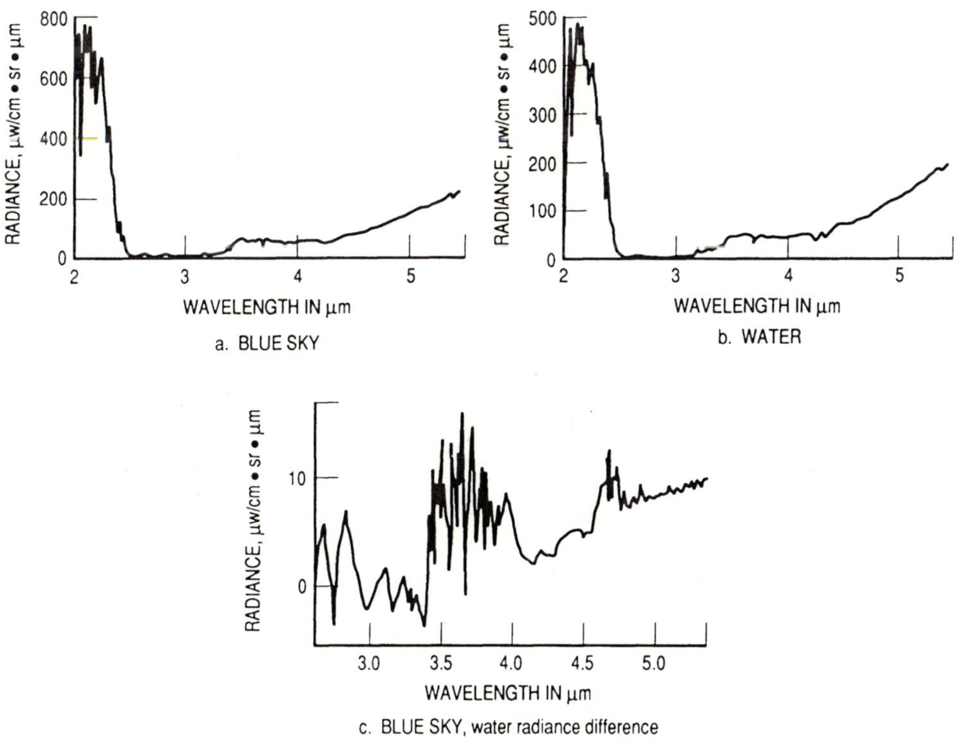

Figure 4.11 Blue sky, water radiance, and radiance difference. (Note change of scale in (c).)

Example

Calculate the maximum clear sky and sunlit cloudy background power as seen by a 50 μr detector with an optical lens of 10 cm. Calculate this value at 1 μm wavelength.

From Figure 4.1 the spectral radiance at 1 μm is

$$9 \times 10^8\ \mu\text{W/cm}^2\ \text{sr}\ \mu\text{m (6000 K sun)}$$

$$8 \times 10^3\ \mu\text{W/cm}^2\ \text{sr}\ \mu\text{m (sunlit clouds)}$$

Using these values in eq. (4.6) for a "bandwidth" of 1 μm and a sunny background, we get

$$h_b = \frac{\pi(10)^2}{4}(50 \times 10^{-6})^2(9 \times 10^8)$$

$$= 176.71\ \mu\text{W} \tag{4.9}$$

$$= 176.71 \times 10^{-6}\ \text{W}$$

$$= -37.52\ \text{dB}$$

Similarly, for a sunlit cloudy background, we have

$$h_b = \frac{\pi(10)^2}{4}(50 \times 10^{-6})^2\ (8 \times 10^3)$$

$$= 0.00157\ \mu\text{W} \tag{4.10}$$

$$= -88.4\ \text{dB}$$

For the purposes of comparison calculate the noise level of the same detector using the formula

$$N = [\alpha_d[(f/\#)D]^2 \Delta f]^{1/2}/D^* \tag{4.11}$$

where N is the noise level, $f/\#$ is the focal number, Δf is the frequency bandwidth of the system, and D^* is the detector detectivity. This equation will be discussed in future chapters, but for now assume the following values:

$$f/\# = 1.5$$

$$\Delta f = 20\ \text{Hz} \tag{4.12}$$

$$D^* = 10^{11}\ \text{cm Hz}^{1/2}/\text{W}$$

Using these values we get

$$N = [(50 \times 10^{-6})^2 (1.5 \times 10)^2 (20)]^{1/2}/10^{11}$$

$$= 3.354 \times 10^{-14} \text{ W} \tag{4.13}$$

$$= -134.7 \text{ dB}$$

From this example, it is clearly seen that for both a sunny and sunlit cloudy background power as seen by the detector is considerably greater than detector noise level. For this reason, the signal strength of the target should compete with background signal h_b for detection. Note that the background radiation was computed for a 1 μm bandwidth. In a narrower or wider bandwidth the background level can be obtained by integration over the appropriate bandwidth.

4.5 GENERATION OF A BACKGROUND SPATIAL RADIANCE MODEL

A credible background spatial radiance model and its statistics can be obtained from the two-dimensional measurement of background radiance, measured for every infinitesimal area of the background in a narrow spectral band and at close range. When such a two-dimensional model is obtained, the background irradiance at the sensor entrance aperture and any desired range and its statistical properties can be calculated if the atmospheric condition is known. In this case, sensor performance parameters, such as the false alarm rate, target detection probability, and any sensor performance parameter influenced by background, can be evaluated.

Although it is impractical to obtain such a two-dimensional radiance function under all practical viewing geometry and atmospheric conditions, it is necessary to make some well-planned measurements of any particular background under most practical conditions.

The background spatial radiance normally is measured by scanning a single detector or array of detectors and recording and processing the resultant output. These measurements are made remotely with a sensor that has a finite detector size, a finite optical image size, and some processing electronics. The measured background radiance is, therefore, the actual background radiance modified by atmospheric transmission, and detector-optics temporal responses.

The following sections discuss the type of statistics governing the background spatial radiance and the limitations on the statistical data obtained from the measurements.

4.6 BACKGROUND STATISTICS

Two types of statistics govern the characteristics of background spatial radiance. The first is related to the types and geometry of background objects such as clouds, the sea, and terrain features like cities, forests, and deserts. The extension and separation of each object in the background influences the first class of statistics, for example, cloud size, separation, and altitudes, influences the statistics on cloud background. The second type of statistics describe the background radiance within each background object, for example, the mean radiance, variance, or peak-to-peak amplitude in a particular cloud system.

The first type of statistic is governed by the background features and its basic components, including trees, buildings, clouds, rocks, and water. The second type of statistics is governed by parameters such as solar angle, viewing geometry, background temperature, and atmospheric absorption path.

For a scanning optical system, the spatial structure of the background appears as the combination of these two statistics, influenced by the sensor transfer function of noise. Because of the effects of large numbers of parameters, we can see why the description of a statistical model of the background radiance is so difficult.

4.7 LIMITATIONS ON MEASUREMENT DATA

Four factors influence the measurement of background radiance data and its statistics: measurement geometry, spectral bands, sensor transfer functions, and sensor noise.

4.7.1 Measurement Geometry

The sensor-background geometry and its relation to the illuminating source have a strong effect on background spatial radiance. Consider, for example, the effect of sensor-background geometry in cloud radiance function. Assume a sensor is located on the ground looking at the sky and is scanning an elevation sector from nadir to the horizon in the daytime. We further assume that the same cloud system, cumulus clouds, for example, have covered part of the scanned sector from nadir to the horizon. We also assume that the cloud spectral reflection and scattering properties are constant along the scanning path. However, the following apparent background differences can be observed:

1. The cloud's apparent angular size reduces as the sensor scans toward the horizon. This could change the apparent cloud size and, hence, the cloud spatial radiance function.
2. Due to the changes in the atmospheric path, and hence the change in the atmospheric transmission, the apparent spatial structure of background radiance varies from the nadir to the horizon.
3. The cloud background is, in general, a nonstationary random function. This is because the apparent cloud radiance depends on the viewing geometry and solar scattering angle, which, obviously, presents different radiances at the nadir than at the horizon.

4.7.2 Spectral Bands

Background spatial radiance is wavelength dependent; that is, a background signal obtained in one spectral band may be quite different from another spectral band. A series of measurements were made for several backgrounds including sky, forest, and cities in several spectral bands [6]. These measurements were made with a scanning radiometer with an instantaneous field of view of 3 mrad operating in four spectral bands of 2 to 3 μm, 3 to 4 μm, 4 to 5 μm, and 8 to 14 μm. Under partially cloudy conditions four records were obtained almost at the same time. A considerable difference was observed between the output of each detector in different spectral bands. The analysis of the experimental data, reported by these investigators, gives an amplitude statistic that indicates a Gaussian distribution for 8 to 14 μm and a Poisson distribution for the same background in the 2 to 3 μm band. Within the experimental limitations, the experience of a great many investigators has shown, however, an amplitude statistic for few backgrounds that approximates in part by a Gaussian curve and deviates in other parts from this curve.

This writer has analyzed the cloud background data measured in the water vapor absorption band with a spatial resolution of about a mile from a high altitude platform for a cloud extension of 160 mi. The amplitude statistics approximated a Gaussian distribution but with a tail extended much above the Gaussian curve. The exact description of this tail has not been made at this time. It can be attributed, however, to solar reflection from high-altitude cirrus clouds or to the bidirectional reflection of the cloud. Obviously when the background consists of two different types of background components such as forest and cities or partial clouds in a clear sky, the background amplitude distribution is not a Gaussian distribution.

4.7.3 Sensor Response

The effect of sensor response on measured background radiance can be divided into two components of system: transfer function and dynamic range. The system transfer function, generally referred to as *modulation transfer function* (MTF), includes the effects of detector spatial and temporal response, electronics, and optics, and if applicable, display system. MTF is related to degradation of optics as one moves away from the boresight of the optical lens. Figure 4.12 gives a one-dimensional sketch of MTF and its components as a function of spatial frequency. As will be seen in Chapter 6, spatial frequency is related to the diffraction limit of the lens much the same way as the gain pattern of an antenna in a microwave system is related to off-boresight angle.

Another parameter of importance in analyzing sensor response is its dynamic range. This can be limited by the dynamic range of the components in the signal flow path. The effective sensor dynamic range is determined by the ratio of the minimum upper signal limit to the maximum lower signal limit.

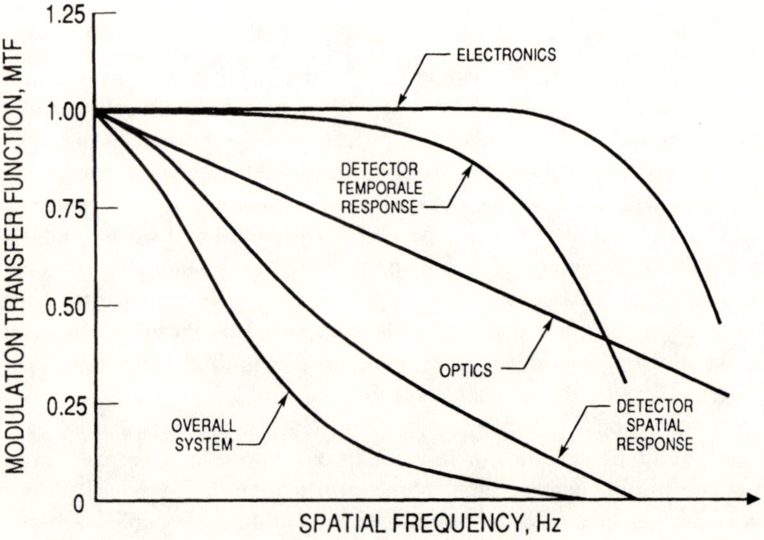

Figure 4.12 MTF of individual components and the overall MTF of a sensor.

The preceding discussion indicates that the measured background data are valid if the magnitude of the output signal is within the dynamic range of the sensor, otherwise the data are distorted by the limitation of the dynamic range of at least one component.

4.7.4 Sensor Noise

The internal noise of the sensor affects the background spatial radiance function and its statistical properties. The mean radiance, autocorrelation, spectral contents, and so on, will be affected by this noise. The question arises as to how the detector noise power and background noise power are added. An experimental method of calibrating the receiver involves the introduction of a known signal into the detector and subsequent measurement of power levels. Using this calibration curve, the background noise level under other conditions can be determined.

4.8 BACKGROUND SPATIAL CORRELATION IN TWO OR MORE SPECTRAL BANDS

When spatial background radiance is measured by an optical sensor in two or more spectral bands, one would expect to sense the two radiances proportionally; that is, the ratio of background radiance in two spectral bands should remain relatively constant as long as observation is made on the same type of background. However, experience has shown differently. In many instances, little correlation can be found between the radiance function measured in one band relative to another as we scan a typical background. This is particularly true around the atmospheric absorption band.

Experimental investigations of cloud background in the absorption bands indicated that, although there were good correlations in the absolute (average) radiance in two spectral bands, high-frequency correlations between the two bands did not actually exist.

If a spectral comparison technique is implemented in an IR system to compare the radiance measured in two or more spectral bands, we should be aware of the type of measurements made as well as the manner of the spectral comparison. To clarify this point, let us analyze the types of spectral measurement made in the field.

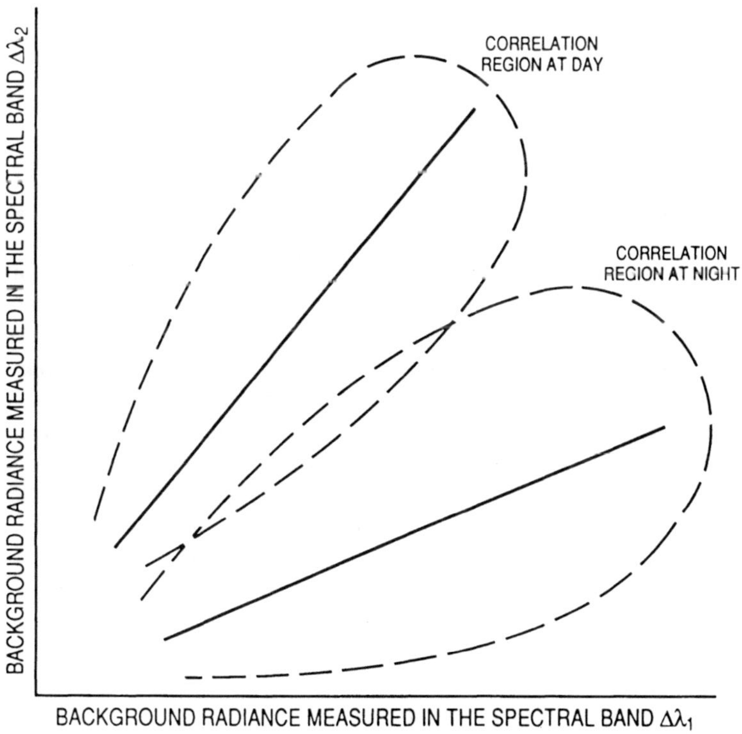

Figure 4.13 Correlation regions for background radiance measured at daytime and nighttime, two spectral bands.

Spectral measurement can be classified into two types:

1. A static measurement of a particular type of background in two or more spectral bands by a nonscanning sensor.
2. A dynamic measurement of the background by a scanning sensor. To obtain all the relevant background data from this type of measurement requires quite an involved effort and is probably one of the most difficult experiments.

In the first case, the background measurement is made by measuring the spectral radiance $N(\lambda)$ of a particular type of background in a continuous spectral band or several discrete bands. The results of these measurements are usually used for spectral correlation to distinguish between a desired object and the undesired ones.

A correlation region of background measured in two spectral bands $\Delta\lambda_1$ and $\Delta\lambda_2$ for day and night is shown in Figure 4.13. The spread around the center line is due to the statistical variation of background radiance caused by a large number of variables as discussed earlier. When multiple observation bands are assumed, the correlation region becomes an n-dimensional region within an n-dimensional space. Each dimension of the space corresponds to the radiance measured in one spectral band.

4.9 TEMPORAL VARIATIONS IN BACKGROUND

Temporal variation of the background can result simply from the fluctuation in the stream of photons expressed by the Bose-Einstein statistics or be caused by the changes in the illumination angle or the motion of the background itself. Examples are moving clouds or changes in the atmospheric constituents in the sensor-background path, such as evolving fog, day and night variations, as well as seasonal variations.

In an atmospheric quantification experiment, one of us and several researchers at the Electro-Optical Research Company have measured the atmospheric transmission in several bands in a 1,000-ft long fog chamber in Richmond, California. This chamber had a provision of injecting fog along the tunnel through several nozzles located at various places along the length as well as the height of the tunnel to provide a uniform fog. When these fog injection processes were stopped, the transmission, as well as particle size and concentration, was measured continuously as the fog decayed.

Figure 4.14 shows the particle concentration versus its diameter for several times after fog injection stopped. Curve *b* indicates the particle distribution 4 s after the cessation of fog injection. The measurement showed a transmission factor for a sensor operating in the spectral band of 8 to 12 μm of about 20%. This transmission reduces to 4% for a sensor operating at 0.95 μm.

Curve *c* of Figure 4.14 shows the particle distribution 40 s after fog injection was stopped. The corresponding transmission data indicates a 92% transmission for

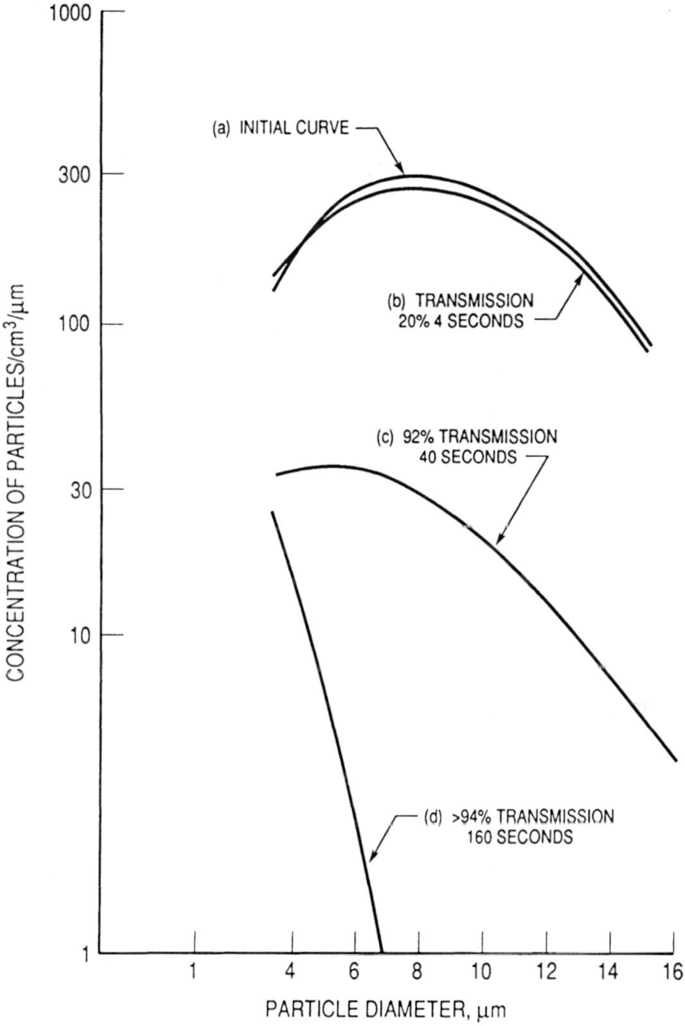

Figure 4.14 Particle concentration and transmission factor when fog starts to decay.

an 8 to 12 μm band, and 58% for 0.95 μm. Curve *d* shows the particle distribution at 160 s after fog decay; no transmission experiments were done at that time.

Example

Compute the background signal level of 300 K blackbody radiation for a laser receiver system with the following parameters [7]:

Collecting lens diameter	10 cm
Detector field of view	1 mr
Optical transmission	1.0
Laser wavelength	10.6 μm
Spectral filter	1% of wavelength

Using eq. (4.6) the background signal level will be

$$h_b = \frac{\pi(10)^2}{4}(0.001)^2(10^{-3})(10.6)(0.01)$$

$$= 8.32 \times 10^{-6}\ \text{W} \tag{4.14}$$

$$= -50.79\ \text{dB}$$

where from Figure 4.1 the value of background radiance of 10^{-3} at 10.6 μm is used.

Assuming a heterodyne laser, the noise level of the system is given by the following equation (see Chapter 8):

$$N = \left(\frac{hf}{\eta}\right)B \tag{4.15}$$

where

- N is the noise level in Watts,
- h is Planck's constant (= 6.623 E − 34 W/s^2),
- f is the laser frequency in Hz,
- η is the quantum efficiency,
- B is the receiver bandwidth.

The receiver bandwidth can be computed from spectral bandwidth of 1% of wavelength or 0.106 μm, and the equation

$$f = \frac{c}{\lambda} \tag{4.16}$$

where c is the velocity of propagation and λ is the wavelength. Differentiating the preceding equation we get

$$\Delta f = B = \frac{-c\Delta\lambda}{\lambda^2} \tag{4.17}$$

Substituting the value of B in the noise equation, we have

$$\begin{aligned} N &= \frac{hc^2\Delta\lambda}{\eta\lambda^3} \\ &= \frac{(6.623E - 34)\,(3 \times 10^{14})^2\,(0.106)}{(0.5)(10.6)^3} \qquad (4.18) \\ &= 9.98E - 9 \\ &= -80.0\ \text{dB} \end{aligned}$$

where a quantum efficiency η of 0.5 is used. From this equation we see that the noise level is over 30 dB below the background level. This means that the background level will dominate the detection process. Note that the diffraction limit of the lens of this laser is

$$\begin{aligned} \theta_d &= \frac{2\lambda}{D} \\ &= \frac{2 \times 10.6E - 6}{10E - 2} \\ &= 212\ \mu\text{rad} \end{aligned}$$

where θ_d is the null-to-null diffraction limit, λ is the wavelength, and D is the diameter of the optics. The detector field of view, however, is one mrad or 1,000 μrad. If the detector field of view is matched to the diffraction limit of the lens, then the background signal level of eq. (4.14) will be reduced by a factor of

$$\begin{aligned} \left(\frac{1000}{200}\right)^2 &= 25 \\ &= 14\ \text{dB} \end{aligned}$$

to a value of ~ 64.79 dB, again exceeding the noise level, but by a smaller amount.

Example

Consider an 8 to 12 μm imaging system with the following parameters:

Collecting lens diameter	20 cm
Detector field of view	100 μr
Optical transmission	1.0

Calculate the average signal level from a sunlit background using data of Figure 4.3. The signal level under the curve of Figure 4.3 for the 8 to 12 μm region can be estimated as

$$(900 \times 10^{-6})\,(12 - 8) = 3.6E - 3\ \text{W/cm}^2\text{-sr}$$

The detector signal level from eq. (4.6) can be written as

$$h_b = \frac{\pi(20)^2}{4}(100E - 6)^2\,(3.6E - 3)$$

$$= 1.130E - 8\ \text{W} \tag{4.19}$$

$$= -70.46\ \text{dB}$$

For the purposes of illustration consider the following noise level equations:

$$\begin{aligned} N &= (A_d\,\Delta f)^{1/2}/D^* \\ A_d &= \alpha_d[(f/\#)D]^2 \end{aligned} \tag{4.20}$$

where

N is the noise level in W,
A_d is the area of the detector,
$f/\#$ is the focal number of the lens,
α_d is the detector field of view,
D^* is the detector detectivity.

Using the following parameters

$$f/\# = 1.5$$

$$D^* = 10^{10}\ \text{cm-Hz}^{1/2}/\text{W}$$

$$\Delta f = 20\ \text{Hz}$$

with the previously given values we have

$$N = [(100E - 6)^2[(1.5)(20)]^2(20)]^{1/2}/10^{10}$$
$$= 1.34E - 12$$
$$= -118.72 \text{ dB}$$

Comparing the preceding noise level to the previously given background signal level of about −80 dB we see a signal-to-noise ratio of about 39 dB. Note that these values do not include atmospheric attenuation, the calculation of which was presented in Chapter 3.

The detection range of the sensor will depend on the desired resolution. To calculate approximate detection ranges consider the following.

The diffraction limit of the lens for the center "frequency" of 10 μm can be calculated by

$$\theta_d = \frac{2.44\ \lambda}{D} \tag{4.21}$$

where θ_d is the null-to-null diffraction limit, λ is the wavelength, and D is the lens diameter. In a great number of cases this equation is approximated by

$$\theta_d \approx \frac{2\lambda}{D} \tag{4.22}$$

for the null to null diffraction limit. Using eq. (4.22) and a value of D of 20 cm we get

$$\theta_d = \frac{(2)(10E - 6)}{20E - 2} \tag{4.23}$$
$$= 100\ \mu\text{rad}$$

which is the same as the detector field of view. Using this value and assuming that we desire a resolution of ΔR of 1 m from Figure 4.15, we have

$$R = \frac{\Delta R}{\theta_d}$$
$$= \frac{1}{100E - 6} \tag{4.24}$$
$$= 10 \text{ km}$$

Thus, higher range values can be achieved if the resolution requirement is relaxed.

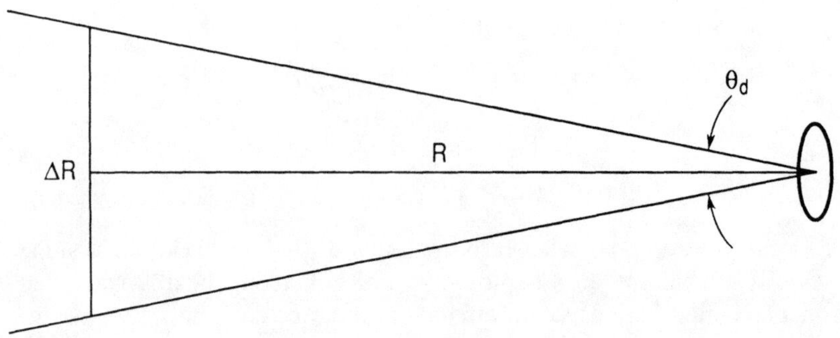

Figure 4.15 Resolution and range dependence in the example problem.

The atmospheric attenuation will also reduce the calculated signal-to-noise ratio. For an atmospheric attenuation of 0.8 dB per km the signal to noise ratio at 10 km will be reduced by 8 dB to a value of

$$39 - 8 = 31 \text{ dB} \tag{4.25}$$

Most of the material covered in the example problems of this chapter will be elaborated in the discussions of future chapters.

REFERENCES

[1] Wolf, W.L., and G.J. Zissis (Ed.), *The Infrared Handbook*, Washington, DC: Office of Naval Research, Department of the Navy, 1985, pp. 3–71.

[2] Bell, E.E., et al., "Spectral Radiance of Sky and Terrain at Wavelengths Between 1 and 20 Microns," *J. Opt. Soc. Amer.*, Vol. 50, December 1960, pp. 1313–1320.

[3] Rose, H., et al., *The Handbook of Albedo and Thermal Earthshine*, Report No. 190201-1-T, Ann Arbor: Environmental Research Institute of Michigan, 1973.

[4] Bell, E.E., et al., "Spectral Radiance of Sky and Terrain at Wavelengths Between 1 and 20 μm," *J. Opt. Soc. Amer.*, Vol. 52, No. 2, February 1962, pp. 201–209.

[5] Earth Observing System Reports, "From Pattern to Process: The Strategy of the Earth Observing System," Vol. 2, NASA, 1988.

[6] Yasumasa, et al., "Statistical Properties of the Background Noise for the Atmospheric Windows in the Intermediate Infrared Region," *Infrared Physics*, Vol. 14, 1974, pp. 17–29.

[7] Hovanessian, S.A., *Introduction to Sensor Systems*, Norwood, MA: Artech House, 1988.

[8] Johnson, R.B., and W.L. Wolfe, (Eds.), *Selected Papers on Infrared, SPIE*, Vol. 513, Parts 1 and 2, 1985.

[9] HIRIS, "High Resolution Imaging Spectrometer," NASA Publication, 1988.

[10] Spiro, I.J., and N. Schlessinger, *Infrared Technology Fundamentals*, New York: Marcel Dekker, 1989.

Chapter 5
Detection and Discrimination in Electro-Optical Sensors

5.1 INTRODUCTION

Optical sensors are of interest in missions where an object with a specified radiance function is to be detected among many similar objects or when it is embedded in a cluttered background. The targets of interest are usually human-made objects whose spectral, spatial, and temporal radiance characteristics are known. Clutter background characteristics, on the other hand, may not be so well defined. Even if they are measured for a particular scenario, they may change for another scenario. Sunlit clouds, terrain features, and sea radiance are examples of these backgrounds.

Detection sensors must incorporate techniques for discrimination of undesired objects and clutter background while enhancing the opportunity for target detection. Unfortunately, the optical characteristics of the target and the background are usually similar. Because of this similarity, a single technique for target detection is not always effective. In most cases it is necessary to employ a combination of several techniques for effective detection and discrimination, and in some instances there is no way to discriminate among similar objects. The discrimination techniques in optical sensors include spectral, spatial, and temporal approaches.

Spectral discrimination is based on the effective target and background radiance differences in several wavelengths. The desired optimization process maximizes the ratio of target irradiance over background irradiance at the sensor focal planes. Its implementation is usually accomplished by selecting proper optical elements, including a spectral filter tailored for this purpose.

Spatial discrimination is based on the differences among the target, similar objects, and background spatial radiance functions. Signal processing, which includes the effects of the optics point spread function, detector size, and its spatial distribution at the focal plane, can be implemented in the processing electronics to maximize the target-to-background spatial radiance differences.

Temporal discrimination is based on the differences in the target and background radiance variation in time and the target motion. The objective of temporal discrimination is to design a circuit that maximizes the signal peak amplitude while reducing background and system noise. A typical detection and discrimination process includes implementation of all of the preceding in a concerted effort.

In its simplified form, an optical detection system consists of collecting optics, an optical filter, a detector, an amplifier, an electronic filter, and a threshold circuit. The radiation from an object is collected, optically filtered, and focused by the optics on a detector that produces an electrical signal proportional to the incident power. This signal, after amplification and filtering, is fed to a threshold circuit. A threshold level based on *a priori* knowledge of the target amplitude and sensor noise establishes the detection criteria of the target.

Coverage of the *field of view* (FOV) is accomplished either by scanning or by staring at the field. Scanning may be done either by a single detector, by multidetectors, or by an array of detectors to scan the FOV in azimuth or elevation or both. Staring is accomplished by a two-dimensional mosaic detector covering the total FOV. Each detector samples the field continuously. The scanning approach has been a classical one since the 1950s. The staring technique is relatively new and has been realized only recently by advances both in hybrid and monolithic focal plane array detectors and in signal processing at the focal plane.

Figure 5.1 illustrates the general scheme of the detection and discrimination process. Input photons are collected by some optical system of refractive or reflective elements. A spectral filter is included within the optical elements to enhance the target signal while reducing noise from background radiation. The photons detected by detectors at the optics focal plane are converted to electrical signals for subsequent processing. The processing includes some electrical filtering for system noise optimization to further enhance target S/N ratio. The resulting signals are then compared with a reference threshold, which is usually some multiple of the noise level, to minimize false alarms caused by noise spikes and pass on signals above some predefined desired minimum signal level. Additional higher order processing, such as multiple threshold exceedance, or multicolor (multispectra) data processing may be undertaken.

The various methods in covering the FOV have actually resulted in different types of sensors. Coverage of the FOV can be achieved through two distinct methods: staring sensors, where the focal plane is populated with thousands of detectors with minimum scanning, or scanning sensors, where the actual optical sensor assembly rotates around a fixed axis or has some of its internal optical elements, lenses or mirrors, perform an oscillatory motion. Normally the elevation direction can be scanned by one optical element and the azimuthal direction by another. In some instances both scans can be accomplished by a single element.

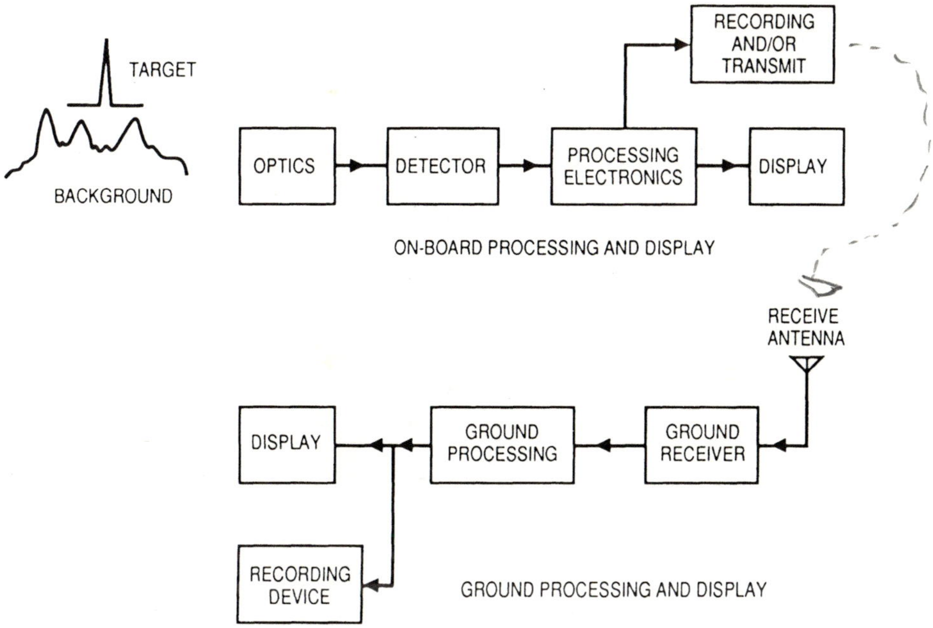

Figure 5.1 General schematic of detection and discrimination process.

5.2 ANALYSIS OF THE DETECTION PROCESS

In detection of a target by an optical sensor it is implicitly assumed that the desired object is moving within the FOV of the sensor with an unknown velocity. The detector scan velocity in a scanning sensor is assumed to be much higher than the target velocity. Therefore the signal pulse is determined by the detector scan velocity. In a staring sensor, however, detection depends on target motion. The signal pulse is therefore determined by the target motion.

In most scanning sensors the detector length is equal to the detector width. In some instances, however, the detector length is several times the detector width. In staring sensors, the width and the length are usually the same.

5.2.1 Signal Amplitude

In calculating the signal amplitude in a detection system, we start with the spatial effect of both the optics and the detector. The optics' most important function is expressed by its point spread function (PSF), which can, for most wide spectral-band applications, be approximated by a two-dimensional Gaussian function. The

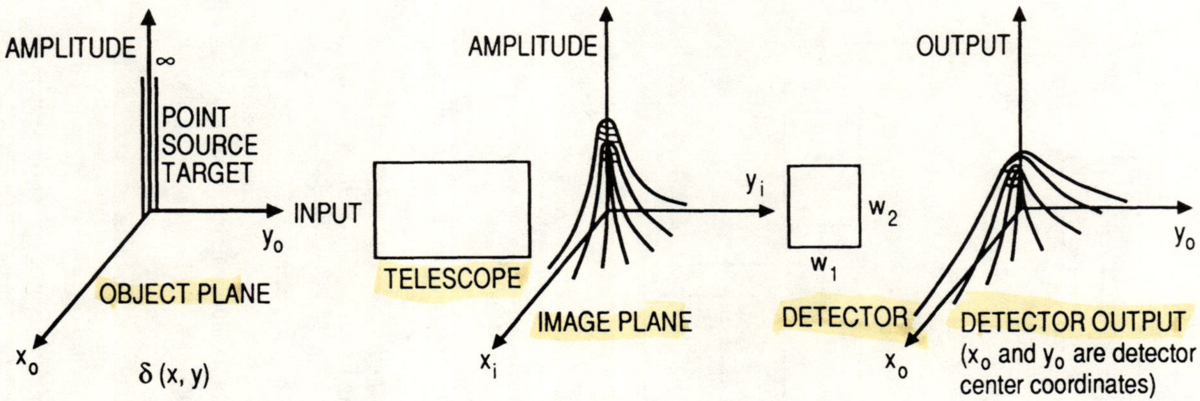

Figure 5.2 Input-output spatial functions of a point source target for a scanning telescope detector.

detector spatial effect, on the other hand, is dictated only by its physical configuration and, unlike optics, is independent of the spectral band of radiation.

Detector size is often chosen to collect at least 90% of the energy of the blur circle whenever the center of the blur circle is on the detector center. To meet this criterion, the standard deviation, σ, of the two-dimensional Gaussian point spread function of the optics is about 23% of the detector width.

A typical PSF of the optics convolved with the detector spatial effect provides a bell-shaped Gaussian signal as a function of position, as shown in Figure 5.2. For a scanning detector, if the target is not at the center of the detector, the amplitude of the signal at the detector may vary as a function of off-center distance. Figure 5.3 gives the relative amplitude of the target for three cases of a target located at the center of a scanning detector and two off-center targets. In a staring detector system the target signal amplitude will depend on its relative position with respect to the detector as illustrated in Figure 5.4.

In a well-designed electro-optical system, the time constant of detection circuitry is matched to the on-target time (or dwell time). This means that all of the energy collected from the target is integrated in the detector circuitry. If the time constant of the detector is not matched to the on-target time, the detector response voltage will be less than the optimum value obtainable.

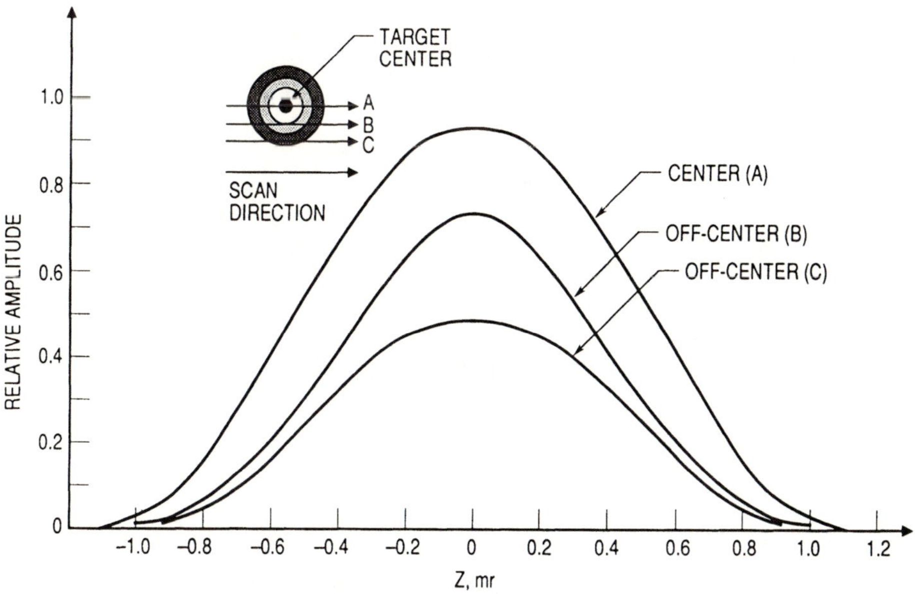

Figure 5.3 Signal amplitude as a function of target location.

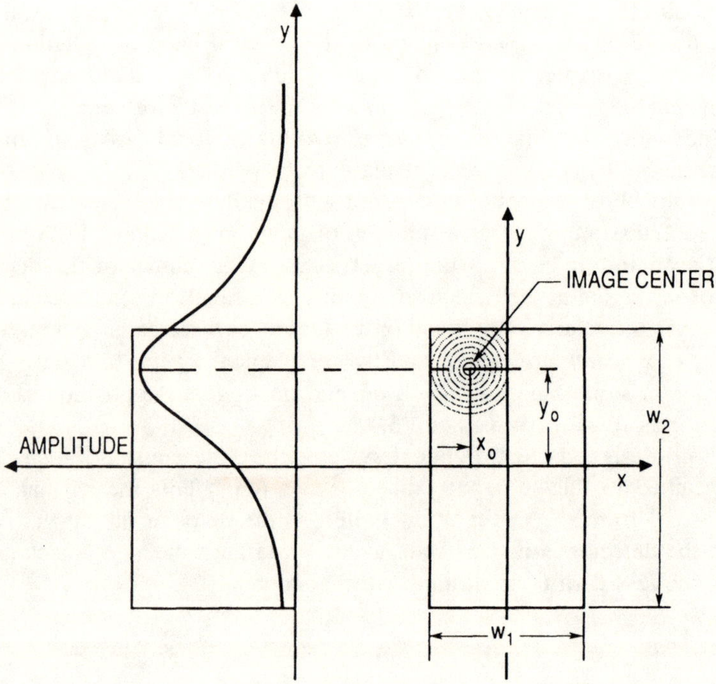

Figure 5.4 Geometry of the irradiance function of a point source target in a rectangular detector.

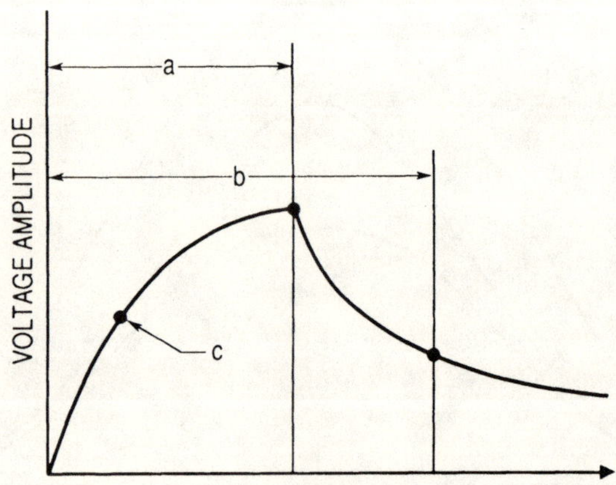

Figure 5.5 Detector time constant and on-target time: detector time constant matched to (a) on-target time, (b) longer than on-target time, (c) shorter than on-target time.

Figure 5.5 illustrates this process starting with the build-up of the voltage amplitude and the drop-off beyond the detector time constant. We see that in both unmatched cases of shorter or longer than on-target time, the target voltage amplitude is lower than its matched value.

In scanning electro-optical systems the dwell time is determined from a consideration of search volume, scan time, and the number of detectors, as will be discussed in Chapter 6. In a staring electro-optical system the dwell time is determined by the number of frames per second; for example, 30 frames per second will give a dwell time of 1/30 s. The dwell time of a staring sensor should also take into account the motion of the target in the FOV. This is similar to the shutter speed of a camera, which is adjusted to the motion of the photographic object.

5.2.2 The Detection Process

Detection of targets in the presence of background (noise) is accomplished by separating the target signal from this background. A typical detection system output, Figure 5.6, shows a signal pulse in the presence of noise. The signal may result from a detector scanning a point source target in the FOV or a staring sensor viewing a flash event such as a laser pulse or a fast-moving object.

The detection process may be implemented by either analog or digital means. Because of the relatively higher cost of digital means, earlier systems used analog filters for detection. With the dramatic advances in digital hardware and the reduction in size, power, and cost, digital filtering techniques are becoming more practical. In the following subsections, both the traditional analog approaches and the digital approaches are discussed.

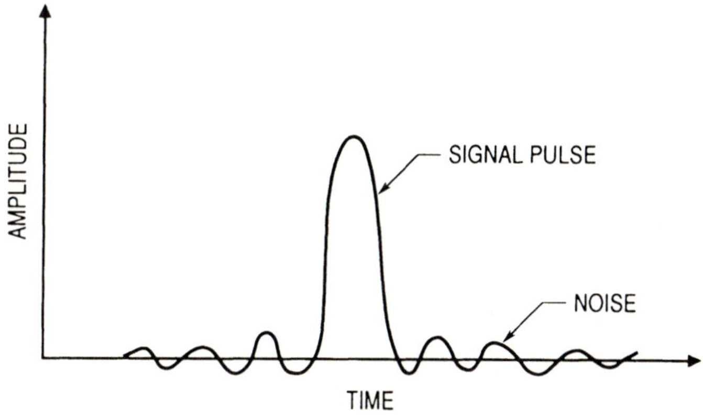

Figure 5.6 Typical pulse signal in noise.

5.2.3 Analog Processing Approach

Selection of an optimum detection filter is influenced not only by detector noise and optics and detector spatial response but also by the background spatial radiance function. Figure 5.7 shows qualitatively the signal and background spectra superimposed on detector noise. Based on this criterion, a typical detection filter takes the shape of a narrow bandpass filter, as shown in Figure 5.8. The lower cutoff is usually influenced by the background spectrum and detector noise, whereas the upper cutoff

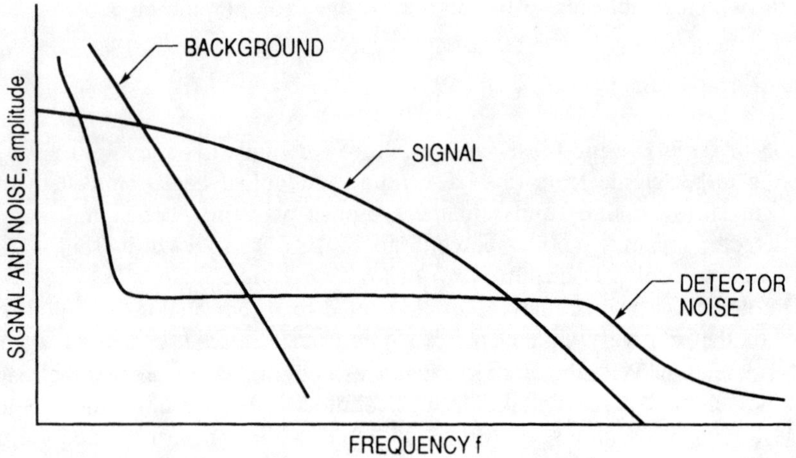

Figure 5.7 Signal, background, and noise power spectrum.

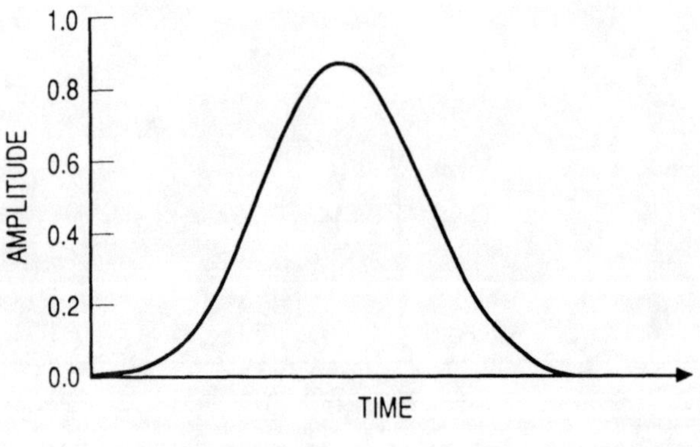

Figure 5.8 Incident power on the detector as a function of time.

is influenced by the detector spatial and temporal response and broadband system noise that dominates the signal at the higher frequencies.

The ideal theoretical analog filter which maximizes peak signal to noise ratio is the matched Weiner filter with the following transfer function:

$$H(j\omega) = \frac{S^*(j\omega)}{N(\omega^2)} e^{-j\omega t_0}$$

where

$H(j\omega)$	is the ideal matched filter transfer function,
$S^*(j\omega)$	is the complex conjugate of the Fourier transform of the input signal, $S(t)$,
$N(\omega)^2$	is the noise power spectrum that includes both detector and background noise.

Although this filter is not realizable, it can be used to compare the response of a practical filter and thus find how closely the practical filter approximates an optimum one. In the case of a scanning detector, the pulse shape is shown in Figure 5.8. The impulse response of an ideal filter as a function of time is shown in Figure 5.9. (Note that in comparing Figures 5.5 with 5.9, the response of Figure 5.5 is the absolute value of the voltage.)

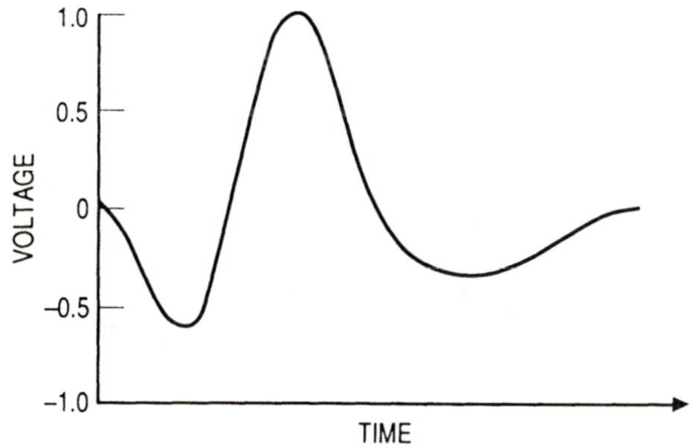

Figure 5.9 Detector response as a function of time.

An important point to consider is that the matched filter does not preserve the input signal wave shape. Detectability of the matched filter is based on the energy in the signal and not on its exact shape. For relatively isolated pulses, the preservation of signal wave shape is usually not important. In some instances, however, the output filter signal wave shape or response may be important and therefore must be considered. Targets may be lost (reduced probability of detection) or false targets may be created (increased false alarms). Figure 5.10(a) shows an example in which a large target pulse is followed closely by a small target pulse. As shown in Figure 5.10(b), the large signal undershoots at its trailing edge. The response of the small signal is superimposed on the large signal undershoot and falls below the detection threshold of the system. The small signal is effectively suppressed by the large signal response.

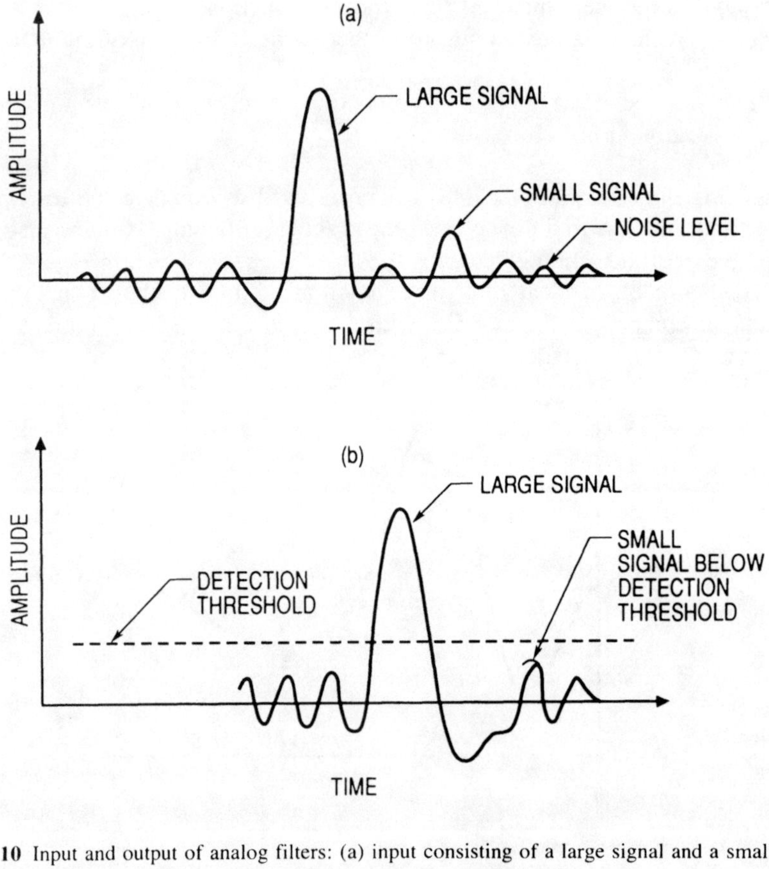

Figure 5.10 Input and output of analog filters: (a) input consisting of a large signal and a small signal; (b) output in which the large signal masks the small signal.

Figure 5.11 shows an example in which a large signal response may generate a false target owing to the slow trailing edge decay.

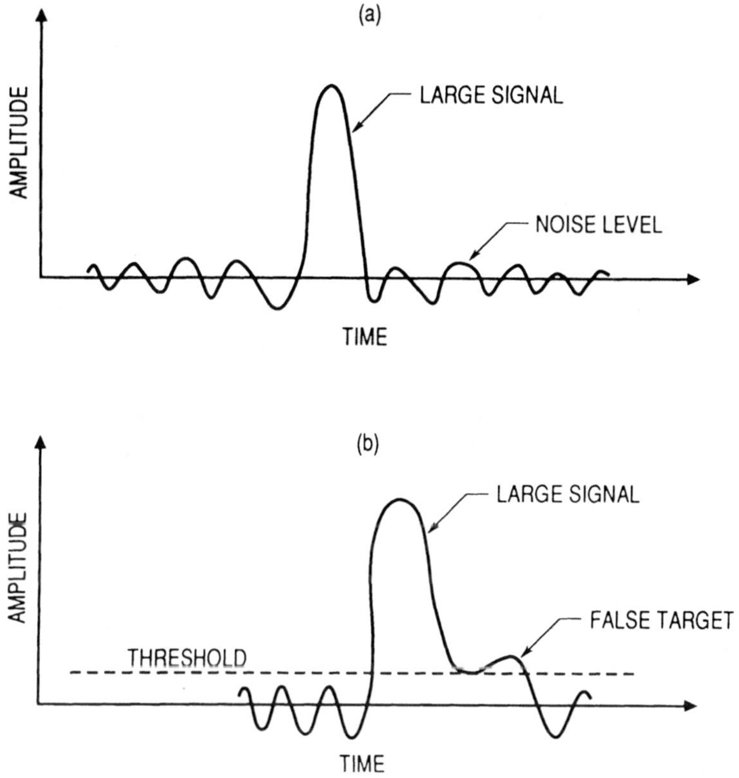

Figure 5.11 False target generated by analog filter response: (a) input signal; (b) filter response.

The response of the filter used for optimizing detection may be tailored to take into account these effects. Because of wide variation in system requirements, each case must be examined individually. Generally, some compromise is necessary to trade off maximum peak signal-to-rms noise and other considerations, as indicated previously. Several standard texts discuss the synthesis of analog filters for various requirements.

5.2.4 Digital Processing Approach

As indicated earlier, the advances in digital hardware now permit the use of digital techniques in the problem of optimal target detection. Within some practical limits (complexity, speed, cost) digital processing may permit more flexibility in the synthesis of optimal filters over purely analog approaches. In addition, digital approaches permit the use of some nonlinear techniques, which may be of benefit.

Relative to the analog approach discussed earlier, a digital filter may permit the synthesis of a more complex and more optimum filter over a purely analog filter constrained by realizability considerations.

An example of the power of nonlinear techniques is illustrated in Figure 5.12. As shown, the output of the digital filter may be "reset" at some point after the passage of the large signal to eliminate the trailing-edge effects of the typical analog-type filter response. Several items are necessary for the implementation of this criterion:

1. The input signal shape characteristics are known. The nominal response therefore may be predicted and an optional reset point may be determined *a priori*.
2. The reset value, based on the long-term "background" level, must be carefully determined. This process implies the use of memory storage and may also require a real-time delay in addition to the nominal filter delay.

It should be apparent from this that the additional digital processing does incur some cost in terms of greater computational complexity.

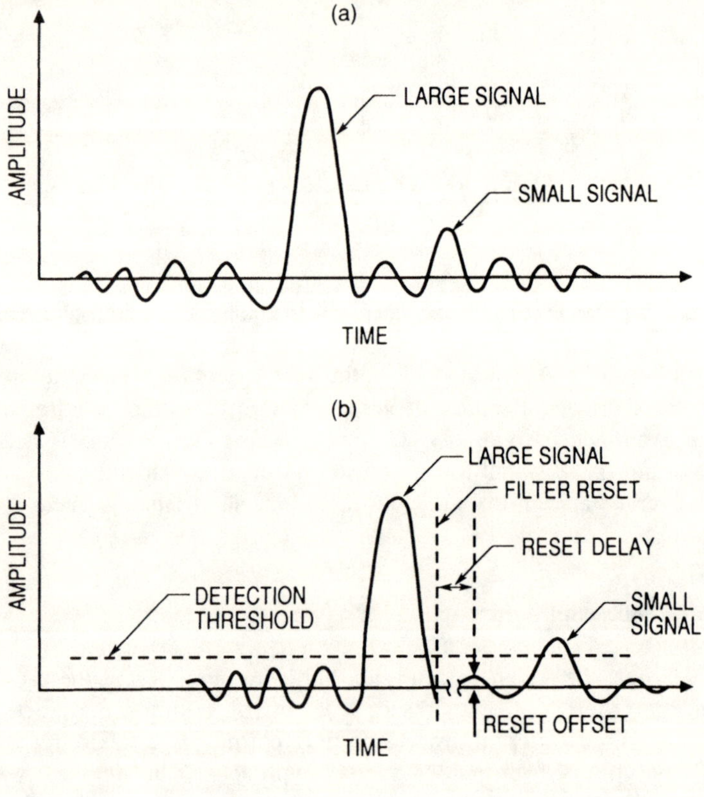

Figure 5.12 Response of a digital filter to consecutive large and small signals: (a) filter input; (b) filter output.

5.2.5 Signal Processing and Staring Mosaic

For a staring mosaic, unlike scanning sensors, the temporal variation of the signal is due to the target motion in the FOV of the sensor. Depending on the relative target motion in the individual detector FOV, the signal output varies as a function of time as well as target location in the detector FOV.

When the detector output is sampled, the output signal amplitude also depends on the relative phasing of the sampling time versus target pulse as it moves across the detector. Several possible frame phasings are shown in Figure 5.13. The amplitude of the target signal is measured as the percentage of the target energy that would be collected in one frame time interval if the target were stationary and located at the position of peak incident power at the detector. The practical problem is that

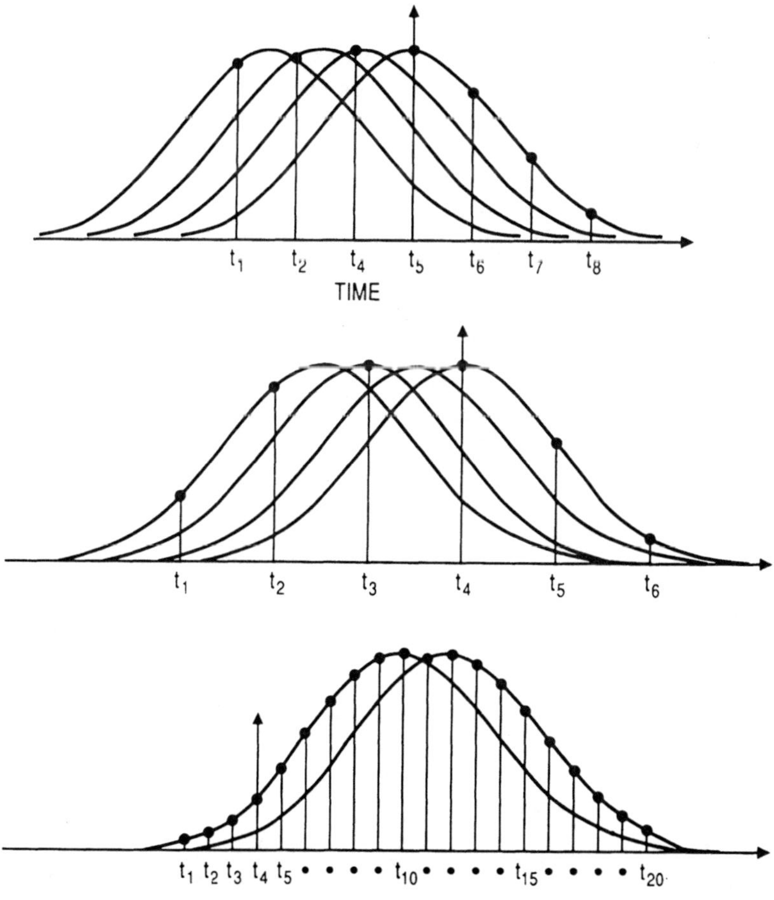

Figure 5.13 Several frame phasing with various sampling ratios.

the relative phasing of the sampling of the detector signal relative to target crossing is unknown.

The performance of a staring array suffers from the large charge collected in the detector element because of background radiation. The goal of the signal processing in this instance is to reject the signal from the dc background and enhance the signal coming from a moving target. The signal processing is designed, therefore, on the basis of a signal differencing technique.

If the charges collected for frame numbers $n - 1$, n, and $n + 1$ are designated by q_{n-1}, q_n, and q_{n+1}, respectively, the first and second differences are defined as

$$\Delta q_n = q_{n-1} - q_n$$

$$\Delta^2 q_n = q_{n-1} - 2q_n + q_{n+1}$$

By applying these equations to the analysis of the pulse interval in Figure 5.13, we obtain the best and the worst phasing of sampling versus target pulse of Figure 5.14. The reason for the rapid rise in this curve is the slow motion of the target, which results in small differences between successive frames. The reason for the small amplitude of fast targets is the decrease in collected energy as the detector dwell time decreases. The worst case target geometry occurs when a target passes between two neighboring detectors along a seam of the array.

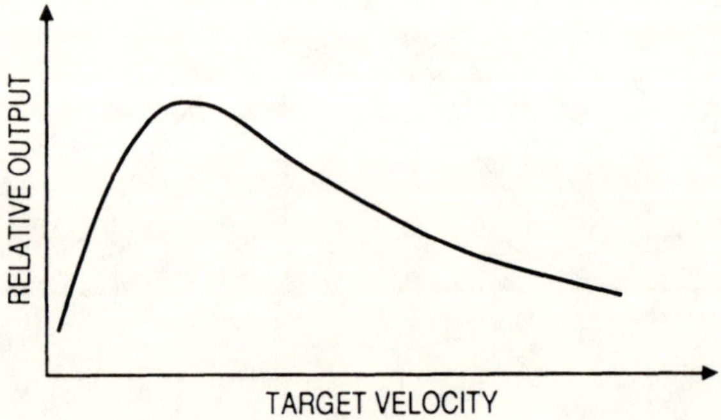

Figure 5.14 Detector relative output versus target velocity.

5.3 DETECTION CRITERIA

The detection of a target signal in the presence of noise is a statistical problem. The satisfactory determination of a target presence may be expressed in terms of the detection probability (the probability of detecting a target when it exists) or the false alarm probability (the probability of a false indication of a target when it does not exist).

The statistical theory of target detection has been treated in several books, including [1, 2]. It is not our intention to give a detailed derivation of this theory, but rather to discuss the significant results of the theory, as applicable to target detection in an optical system.

The single scan detection probability and the multiple scan detection probability are the most commonly used criteria for target detection in an optical search system. The following discussion evaluates these two criteria and the false alarm rate of an EO sensor based on the following assumptions:

1. Noise at the detection filter input has a Gaussian distribution and therefore can be described completely in terms of its rms value. For a system whose performance is limited by detector noise, the assumption of Gaussian noise usually is valid, particularly when the processing filter placed at the output of the detector has a narrow bandpass. This assumption may not be valid, however, for a scanning sensor, when background spatial radiance is significant.
2. The detector output signal produced by a target is a single pulse described by its peak value.

The conventional technique for detection of a target signal in an optical sensor involves processing the amplified detector output signal through a detection filter. The filter output is then fed into a linear thresholding circuit, the level of which is either fixed or adaptive. Preset fixed thresholds are based on *a priori* knowledge of the target radiance level and the rms value of the system noise. Adaptive thresholds are set based on the current noise level of the system, which is continuously measured. If the amplitude of the output of the filter is above the threshold level, a detection is made. If the signal is present, it is a true detection. If the signal is not present, the detection is a false alarm.

5.4 PROBABILITY OF DETECTION

The theoretical detection model used in electro-optical systems is similar to the models used in radar and laser systems and the detection of sinusoidal signal in Gaussian noise. This model is popular primarily because an analytical solution was obtained where the probability of detection can be related to signal-to-noise ratio. In recent years, however, it has been shown that in a number of cases the background noise arising from cloud formation or Earth reflection may not be Gaussian. In these cases the Gaussian model described here can be used for preliminary studies. The refinements of this model for non-Gaussian situations can be performed through computer simulations.

Consider a sinusoidal signal augmented by Gaussian noise in the detection filter. The expression for the probability of detection of a target having a signal-to-noise power ratio of S/N is given by [3]

$$P_s = 1 - \int_0^{E_t} \nu \exp\left[\frac{-\nu^2 - 2\frac{S}{N}}{2}\right] I_0\left\{\sqrt{\frac{2S}{N}}\right\} \mathrm{d}\nu \tag{5.1}$$

where

- P_s is the probability of the signal exceeding threshold E_t (probability of detection),
- E_t is the threshold setting, in volts,
- ν is the envelope of signal plus noise,
- S/N is the signal-to-noise power ratio,
- I_0 is the hyperbolic Bessel function of zero order,

and the factor of *two* in the S/N ratio is used to convert the signal to peak signal voltage value. In special cases where no signal is present eq. (5.1) yields the probability P_n of a noise only exceeding threshold voltage E_t:

$$P_n = 1 - \int_0^{E_t} \nu \exp\left(-\frac{\nu^2}{2}\right) = \exp\left(-\frac{E_t^2}{2}\right) \tag{5.2}$$

where E_t is peak signal voltage, $E_t^2/2$ will be threshold-to-noise power ratio.

Equations (5.1) and (5.2) after a number of mathematical substitutions can be evaluated parametrically. Figure 5.15 gives the probability of detection as a function of signal-to-noise ratio with probability of false alarm as a parameter.

Note that eq. (5.2) describes the relation between root mean square power threshold setting $E_t^2/2$ and the probability of false alarm P_n. As an example, for a probability of false alarm $P_n = 10^{-4}$, we have

$$P_n = e^{-E_t^2/2}$$

$$\frac{E_t^2}{2} = 9.2 \tag{5.3}$$

$$= 9.46 \text{ dB}$$

Thus, an rms power threshold setting of 9.46 dB above the noise level will result in a probability of false alarm of 10^{-4}. Note also that this threshold setting from Figure 5.15 corresponds to the 50% probability of detection value on the 10^{-4} false

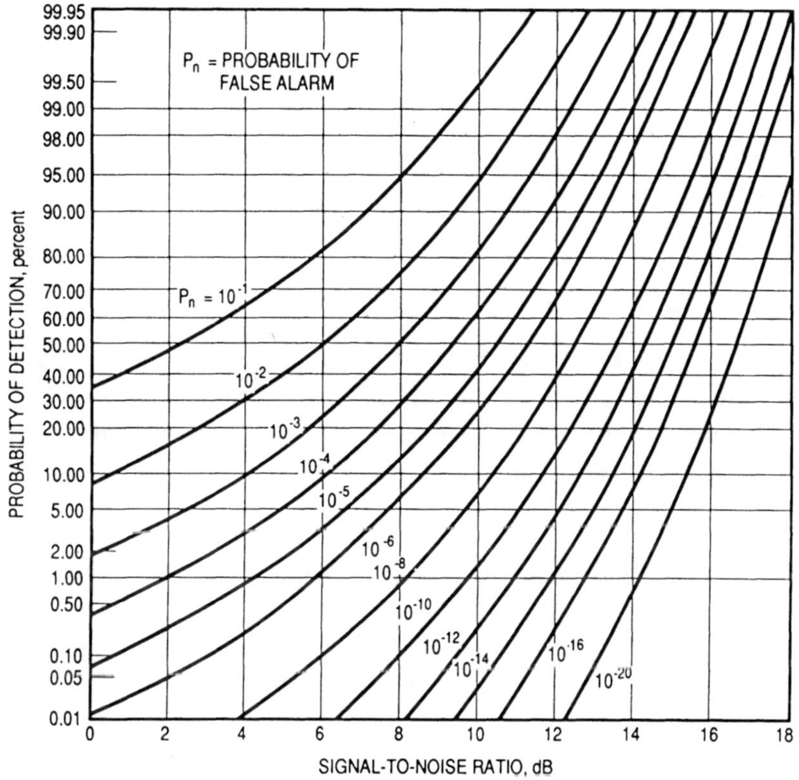

Figure 5.15 Probability of detection as a function of S/N ratio with the probability of false alarm as a parameter.

alarm probability curve. Thus, this may be an easier way of obtaining the threshold value.

The threshold setting of 5.3 is relative to the noise level; this value may be considered as threshold-to-noise ratio.

The detection probability and false alarm rate are often given in system specifications. Generally, the question is, What minimum signal-to-noise ratio will satisfy those requirements? The normal procedure for satisfying this requirement follows.

Example

Consider that a detection filter with a bandwidth of 100 Hz has been selected. For a false alarm rate of one per hour, calculate the S/N ratio required for a 90% detection probability.

The time constant of the filter is inversely proportional to its bandwidth, or

$$T_f = \frac{1}{\Delta f} = 0.01 \text{ s} \tag{5.4}$$

The false alarm rate per second will be $1/3600 = 2.778E - 4$. Thus, the probability of false alarm per filter time constant will be

$$(2.778E - 4)(0.01) = 2.778E - 6$$

The probability of false alarm in each filter time constant will be about $3E - 6$. Using the curves of Figure 5.15, this false alarm probability will result in a threshold-to-noise ratio power setting of 11 dB corresponding to a 50% detection probability and a S/N requirement of 13 dB for 90% probability of detection.

5.4.1 Illustration of Probability of Detection and Probability of False Alarm

Figure 5.16 illustrates the amplitude distributions of noise alone and signal plus noise. The distributions are the well-known Gaussian distributions. The threshold value, E_t, is also shown. With only noise present, there is a finite probability of a noise spike exceeding the threshold and causing a false signal indication or false alarm.

Assuming a Gaussian distribution of noise, the probability of false alarm (P_n) is simply the integral of the distribution or area under the curve to the right of the threshold. When a true target is present, noise added to the target signal may cause the overall composite signal to fall below the threshold. Therefore, the probability of the signal exceeding the threshold, or the probability of detection (P_s), is given by the integral of the curve to the right of the threshold.

The selection of an optimum threshold level is influenced by the preceding criteria and also other important sensor design issues, such as

1. Variations in target radiance due to spectral reflectivity or emissivity or temperature variations of the target.
2. Variations in sensor response due to variation in the optics point spread function, optical off axis, and its spectral response.
3. Variation in target-sensor viewing geometry.

These factors must also be considered in the final selection of an optimum threshold.

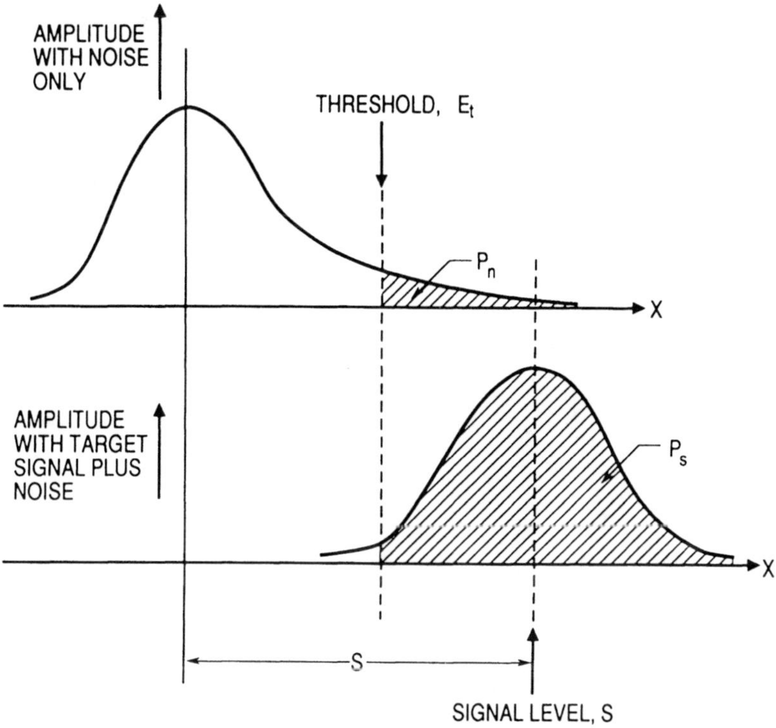

Figure 5.16 Distributions of signal and noise and probabilities of detection and false alarm for a single detector channel.

5.4.2 Multiple Threshold Exceedance Detection

This section discusses detection probability and false alarm rate of an EO sensor and describes the optimum rules or algorithms to use for multiple threshold exceedance detection. Examples of detection in sequential, parallel, and hybrid channel systems are covered. Throughout this discussion we assume that the objective is to detect a target in the presence of white Gaussian noise. Detection is based on the exceedance of a fixed threshold.

Figure 5.17 illustrates the basic elements of a simple threshold exceedance detection system. The detector converts the input incident power into electrical signals that are amplified and filtered. The filter outputs are compared with a threshold value, E_t. A filter output exceeding the threshold is called a *threshold exceedance*. The presence of random additive noise introduces uncertainty into the outcome of the threshold exceedance. This outcome may be either a detection of a target or a false alarm due to noise.

5.4.3 Event Rates and Sample Gates

The rate at which events occur is of concern in the design of the data processing systems that follow the threshold process. The event rates are determined by the filter, $G(f)$, shown in Figure 5.17. The filter restricts the highest frequency components that pass through the system and, therefore, the highest event rate of significance. For noise alone, the average event rate or false alarm rate is given by

$$N_f = \frac{P_n}{T_f} \tag{5.5}$$

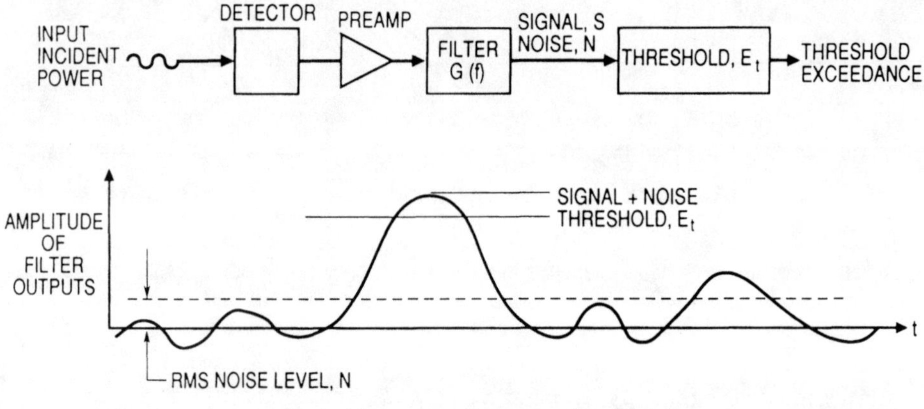

Figure 5.17 Basic EO detection system elements.

where

N_f is the false alarm rate,
P_n is the probability of a single event noise pulse exceeding the threshold,
T_f is the filter time constant.

The filter time constant can be approximated by the inverse of filter bandwidth Δf, or from eq. (5.5) we have

$$N_f = P_n \Delta f \tag{5.6}$$

For example, for a threshold setting of 2.5 dB and a frequency bandwidth of 1000 Hz from Figure 5.15, we have a false alarm probability of 10^{-1}. This corresponds to a S/N ratio of 2.5 dB at 50% detection value. The false alarm probability is

$$N_f = (0.1)(1000)$$

$$= 100 \text{ per second}$$

5.4.4 Multiple Channel Detection

For multiple channels, we would like to compare threshold outputs for simultaneous occurrences. The question of relative phasing of the output exceedances is important. Figure 5.18 illustrates the output from two channels. As shown, the two channels generate very narrow pulses with the output just exceeding the threshold. A direct AND gate would result in a zero output, because of the slight phase differences in output, even though the thresholded output is within a reasonable correlation distance

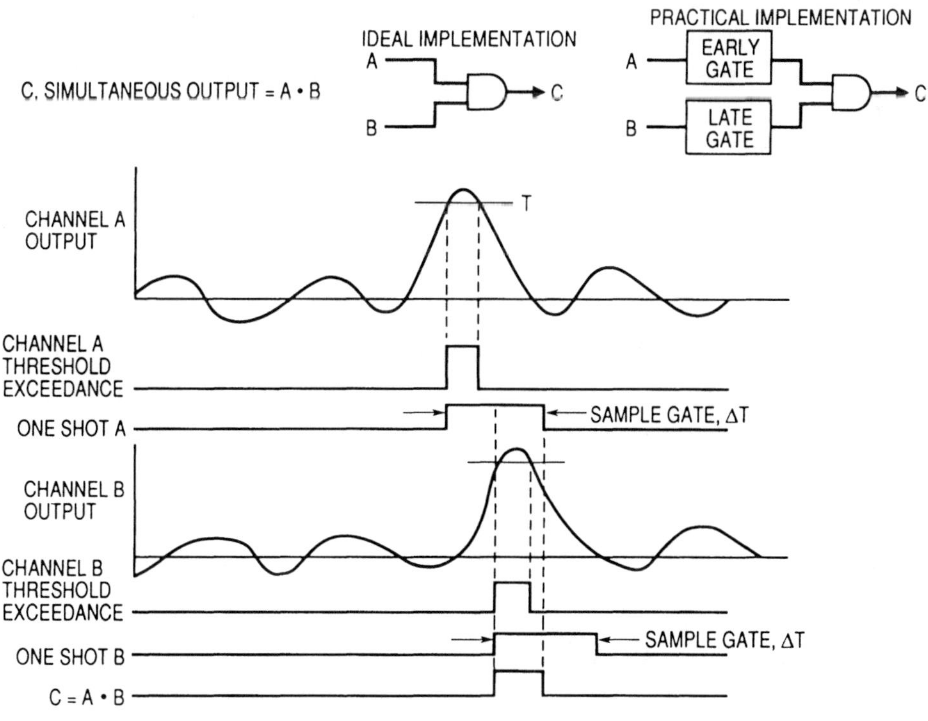

Figure 5.18 Multiple channel detection.

or time of each other. Therefore, a practical implementation as shown may be used. This implementation uses gates of duration ΔT that are delayed between the two channels.

As an example of coincidence detection consider the system of Figure 5.19. This figure shows a space-based system designed to detect very short optical pulses. To discriminate against random pulses (such as gamma pulses), which may look like the optical pulse, two detectors are used in parallel. The optical energy is split 50/50 by means of a partly silvered mirror. To declare a valid input, both detectors must "see" the input simultaneously. An undesired pulse (such as gamma ray) that strikes only a single detector will not pass through the coincidence logic.

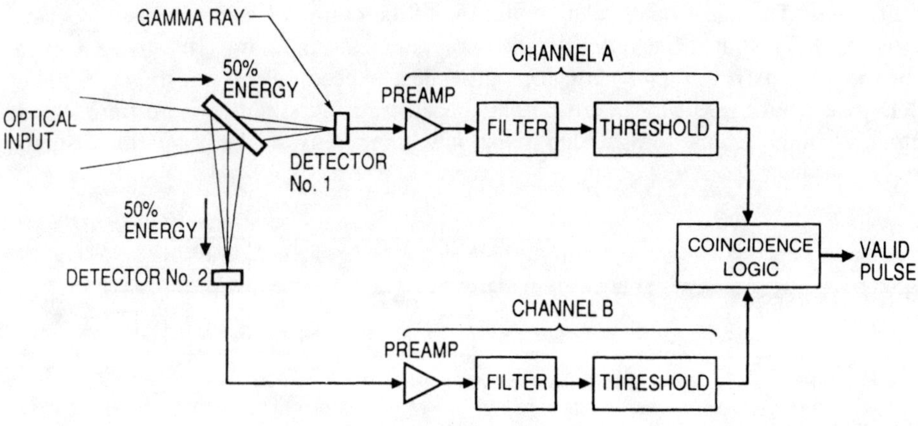

Figure 5.19 Example of coincidence detection system.

Assuming that both channels are identical, the probability of detection and the probability of false alarm are given by the same form:

$$P_D = P_s^2$$

$$P_F = P_n^2$$

Thus, both probability of detection and of false alarm will be considerably reduced from their single channel values.

5.4.5 Multiple Event Occurrence

The probability of multiple events occurring can be used in detection of targets where multiple channels in parallel or in series are used for detection. Consider an event

A with a probability of occurring θ where $0 \leq \theta \leq 1$. We perform v observations and ask for the probability P_r that event A occurs in r, $0 \leq r \leq v$, of these observations:

$$P_r = \binom{v}{r} \theta^r (1 - \theta)^{v-r} \tag{5.7}$$

where

$$\binom{v}{r} = \frac{v(v-1)(v-2)\ldots(v-r+1)}{1, 2, 3, \ldots, r}$$

$$= \frac{v!}{r!\,(v-r)!}$$

and the symbol ! denotes factorial product. This equation for a two out of two event with the probability of event occurring of 0.75 will result in

$$P_r = 0.5625$$

Another important relation with respect to detection is the answer to the question, What is the probability of an event A occurring in at least one of N experiments? Assuming the probability of the event occurring is P_d, the probability of the event not occurring will be $1 - P_d$ and the probability of the event not occurring in N experiments will be

$$(1 - P_d)^N$$

Finally, the probability of the event occurring at least once will be

$$P_1 = 1 - (1 - P_d)^N \tag{5.8}$$

where P_1 is the probability of the event occurring at least once in N experiments.

5.4.6 Sequential Detection

As an example of sequential detection consider the EO scanning system of Figure 5.20, consisting of an array of vertical detectors. Scan waveform, target signal, and detector outputs are given in the figure. Note the active scan and retrace portions of the scan waveform and the reduced target signal on retrace portion. This reduction of the target signal is due to the fast motion of the detector across the target in the retrace period and the resulting reduced target dwell time.

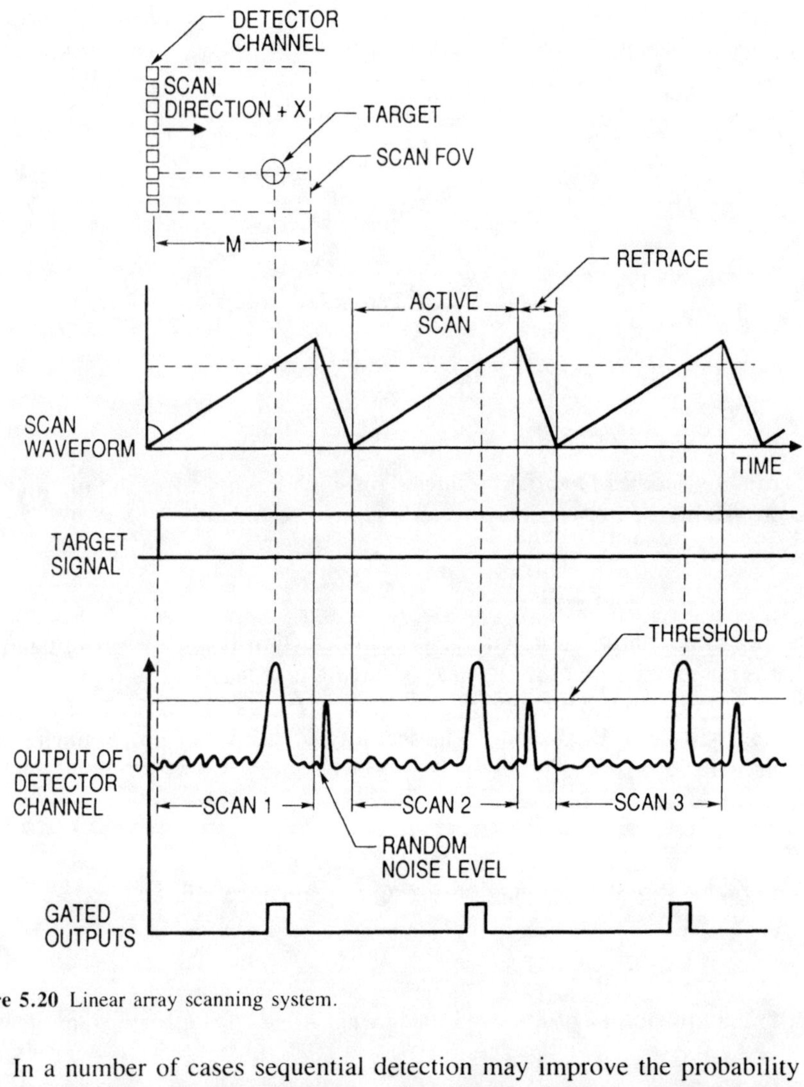

Figure 5.20 Linear array scanning system.

In a number of cases sequential detection may improve the probability of detection and reduce the false alarm rate. Consider the following example. Assume a S/N ratio of 10 dB and a threshold setting of 8 dB. From Figure 5.15 this threshold setting corresponds to a probability of false alarms of 10^{-3} with a S/N ratio of 10 dB representing an 80% probability of detection. Consider the case that in a two out of two sequential detection we require the same probability of false alarm value. The single-look false alarm probability then becomes

$$P_n = (10^{-3})^{1/2} = 0.0316$$

Using this probability of false alarm, from Figure 5.15, for a 10 dB S/N ratio we get a probability of single-look detection of 97%. Using this, the two out of two detection will be

$$P_d = (0.97)^2 = 94\%$$

An improvement on the original detection probability of 80% with the same false alarm rate.

A combination of detection criteria can be used in sequential detection such as two out of three, or three out of five, and so on. The detection and false alarm probabilities can be computed using eqs. (5.7) and (5.8).

It is important to point out that the improved performance does not come free, it requires extra cost as to

1. Greater processing complexity to store frames of data and apply the algorithm.
2. Additional delay in the output decision. Delays are at least equal to the number of frames of data required to apply the rules.

5.4.7 Time Delay Integration

Another method of improving the probability of detection is by the use of time delay integration (TDI), where the signal-to-noise ratio from one detection corresponding to a specific dwell time is added to a signal-to-noise ratio of an adjacent detector. The resulting S/N ratio improvement will be $\sqrt{n}$, where n is the number of sequential detections.

Consider the previously given numerical example with signal- and threshold-to-noise power ratios of

$$\frac{S}{N} = 10 \text{ dB}$$

$$\frac{T}{N} = 8 \text{ dB}$$

and a single-look probability of detection of 80%. It was shown that a sequential detection of two out of two will improve the detection probability to 97%. Using TDI and adding subsequent detections to the previous one, we will realize an improvement of factor of $\sqrt{2}$ (1.5 dB) in S/N ratio. The corresponding detection probability for a S/N ratio of 11.5 dB from Figure 5.15 will be 88%.

Time delay integration may be extended to several integrated detections for improved signal-to-noise ratio and subsequent detection probability.

A note of caution should be given here. The theoretical improvement in S/N ratio is $\sqrt{n}$, where n is the number of pulses integrated. In practice, however, there are many reasons why the theoretical value may not be reached. Among them,

1. The delay value may not match the exact required value due to the scan rate when scanning is done mechanically.
2. The scan may be skewed, which causes an offset in the detectors that increases with n.
3. The mechanism for delaying the analog output and summing them may introduce additional noise, which will degrade the resultant signal-to-noise ratio.
4. The system jitter and instability may cause motion of detectors from one look to the other.

In summary, relative performance for TDI versus multiple thresholding should be evaluated for each EO system, considering the scan type, focal plane arrays, jitter, and so on.

REFERENCES

[1] Seyrafi, K., *Electro Optical System Analysis*, Los Angeles: Electro-Optical Research Corporation, 1985.

[2] Lawson, J. L., and G. E. Uhlenbeck, *Threshold Signals*, New York: Dover Publications, 1950.

[3] Hovanessian, S. A., *Radar System Design and Analysis*, Norwood, MA: Artech House, 1984, p. 139.

[4] Johnson, R. B., and W. C. Wolfe (Eds.), *Selected Papers on Infrared Design, SPIE*, Vol. 513, Part 1 and 2, 1985.

[5] Hoel, P. G., *Introduction to Mathematical Statistics*, New York: John Wiley and Sons, 1966.

Chapter 6
Electro-Optical System Design and Performance Equations

6.1 INTRODUCTION

The electro-optical system design and performance equations together with numerical examples of these equations are given in this chapter. Initially, the signal-to-noise ratio equation is derived for a point source target. Following this, the *noise equivalent flux density* (NEFD) relation is obtained. For the class of distributed targets or background, the signal to noise ratio equation, including the integral of target spectral emittance, is given. The performance criteria depicted by *noise equivalent temperature* (NET) and the *minimum resolvable temperature* (MRT) are also derived and discussed.

Probability of detection curves as a function of signal-to-noise ratio, with the probability of false alarm as a parameter, are given and used in numerical examples.

Image quality considerations relating target detection, orientation, recognition, and identification ranges to MRT values are discussed and illustrated through numerical examples.

6.2 BASIC EQUATIONS

The basic equations in the design and analysis of electro-optical systems, as well as all other sensor systems, involve target signal power and system noise power. As expected, the higher is the signal-to-noise ratio, the greater the probability of target detection for a given threshold setting. And, the higher is the threshold setting, the lower the number of false alarms.

For a point source target radiating in a narrow bandwidth with radiation intensity J_t, the irradiance at the entrance aperture of the optics is given by

$$H_t = \frac{J_t}{R^2} \tag{6.1}$$

where

J_t is the radiation intensity, W/unit solid angle,
H_t is the irradiance at the optics aperture, W/cm^2,
R is the sensor-target range, cm.

Note that eq. (6.1) does not include any atmospheric transmission loss. Atmospheric losses, however, should be included in the final target signal values.

Multiplying eq. (6.1) by the optical aperture area, we get the total input signal power:

$$S = \frac{\pi D^2}{4} \cdot \frac{J_t}{R^2} \tag{6.2}$$

The signal represented by eq. (6.2) is at the entrance aperture of the optics; there will be an optical transmission loss between this point and the detector, as will be discussed later in system losses.

The detector noise level is given by the relation

$$N = (A_d \Delta f)^{1/2} / D^* \tag{6.3}$$

where

N is the noise power level, W,
A_d is the detector area, cm^2,
Δf is the frequency bandwidth of the received signal, Hz,
D^* is the detector detectivity, cm-Hz$^{1/2}$/W.

The terms of the noise power equation (6.3) need additional discussion, as given in the following paragraphs.

For a square detector, the detector area A_d is related to the instantaneous field of view α_d and the focal length f_1 as follows and described in Figure 6.1:

$$\begin{aligned} \text{detector area} &= A_d \\ &= (\alpha_d^{1/2} f_1)(\alpha_d^{1/2} f_1) \\ &= \alpha_d f_1^2 \end{aligned} \tag{6.4}$$

The focal length is related to focal number $(f/\#)$ and aperture diameter by

$$f_1 = (f/\#)D \tag{6.5}$$

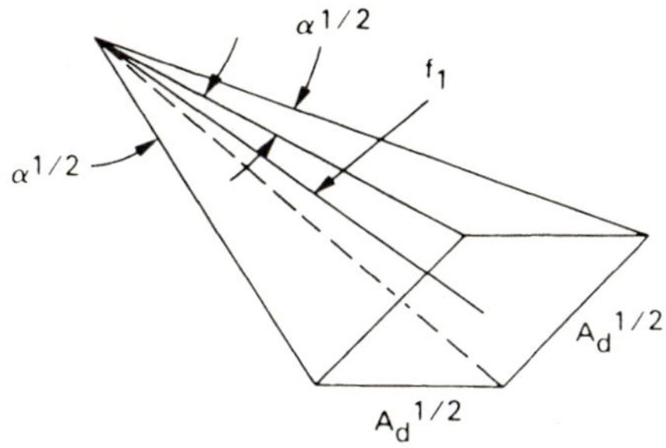

Figure 6.1 Detector area, focal length, and instantaneous field of view.

Using this, the detector area becomes

$$A_d = \alpha_d[(f/\#)D]^2 \tag{6.6}$$

To calculate the next term of the noise power equation, Δf, we need to calculate the detector on-target time or, equivalently, the duration of time during which the detector receives target energy. As we will see later, this time duration can be converted to frequency increment Δf through Fourier transform analysis.

Assuming the total solid angle coverage of the sensor to be Ω_s sr, the number of resolution elements will be this value divided by the instantaneous field of view of each detector:

$$n_r = \frac{\Omega_s}{\alpha_d} \tag{6.7}$$

where

- n_r is the number of resolution elements,
- Ω_s is the total solid angle coverage of the sensor, sr,
- α_d is the instantaneous field of view of each detector, sr.

For a frame time of T_f, the on-target time (dwell time) per resolution cell will be

$$t_d = T_f/n_r \tag{6.8}$$

where

- T_f is the frame time, s,
- t_d is the on-target time, s.

The value of the on-target time given by eq. (6.8) is for a single detector. If the system is equipped with n detectors in parallel, the on-target time will be increased by this factor; that is,

$$t_d = \left[\frac{n\alpha_d T_f}{\Omega_s}\right] \tag{6.9}$$

where n is the number of detectors.

Note that eq. (6.9) does not include scanning efficiency and other losses associated with detector on-target time. These losses will result in a scanning loss factor that can be incorporated in eq. (6.9) as follows:

$$t_d = \frac{n\alpha_d T_f}{\Omega_s} K_n \tag{6.9a}$$

where K_n is the total scanning loss factor.

The relationship between on-target time and the bandwidth of the received signal is obtained from Fourier transform analysis. The Fourier transform of a step function of duration t_d can be expressed by a $\sin x/x$ relationship. This $\sin x/x$ represents the spectrum of the step-function time signal. The approximate half-power bandwidth (3 dB) of this spectrum is $1/t_d$ with the null-to-null bandwidth of $2/t_d$ [1]. Using the fact that the filter bandwidth is inversely proportional to the on-target time with proportionality constant K_{fd} we have

$$\Delta f = \frac{K_{fd}}{t_d} \tag{6.10}$$

where

Δf is the equivalent filter noise bandwidth, Hz,
K_{fd} is the bandwidth proportionality constant.

Using the value of t_d from eq. (6.9) in (6.10) we get

$$\Delta f = \frac{\Omega_s K_{fd}}{n\alpha_d T_f K_n} \tag{6.11}$$

Substituting the parameters of noise power equation (6.3) from eq. (6.6) and (6.11), we get

$$N = \frac{D(f/\#)}{D^*}\left(\frac{\Omega_s}{nT_f}\frac{K_{fd}}{K_n}\right)^{1/2} \tag{6.12}$$

Using eq. (6.12), together with the signal power equation (6.2), we can write the signal-to-noise ratio equation as

$$\frac{S}{N} = \frac{\pi D J_t D^* L}{4(f/\#)R^2}\left(\frac{nT_f}{\Omega_s}\right)^{1/2} \tag{6.13}$$

The loss factor L has been added to the numerator of this equation to account for various system losses. This factor includes

$$\begin{aligned} L &= K_o K_e K_d K_m (K_n/K_{fd})^{1/2} \\ &= r_s (K_n/K_{fd})^{1/2} \end{aligned} \tag{6.14}$$

where

K_o is optics,
K_e is electronics,
K_d is the detector,
K_m is monitor,
r_s is the product of the preceding representing system modulation transfer function, MTF,
K_n is the total scanning loss factor,
K_{fd} is the bandwidth proportionality constant.

The values of coefficients K given in eq. (6.14) usually range between 0.5 and 1.0 with the exception of K_{fd}, which may be greater than unity. The coefficient K_{fd} also may be included in the system MTF value r_s.

Note that the S/N equation (6.13) does not include atmospheric losses, which are range dependent. These losses can be calculated by methods outlined at the end of this chapter.

6.2.1 Detector Detectivity Values

Commercially available detectors with measured sensitivities are used in electro-optical systems. Figure 6.2 [2] gives detectivities for the background limited condition of 300 K and a 180° field of view. The sensitivity of the detector can be increased considerably under limited background conditions, where the detector is shielded to a limited cone angle as shown in Figure 6.3. The theoretical limit of D^* in this case is given in [3] and reproduced here as Figure 6.4. Note that this theoretical limit is 100 times better than the D^* values obtained without a cooled spectral filter.

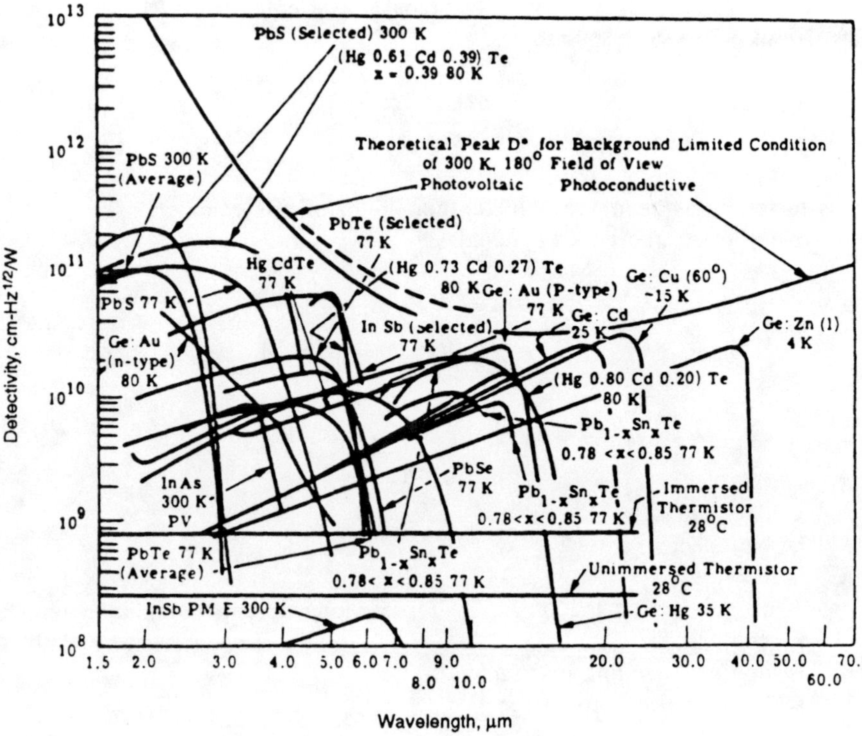

Figure 6.2 Spectral D^* for a number of commercially available detectors.

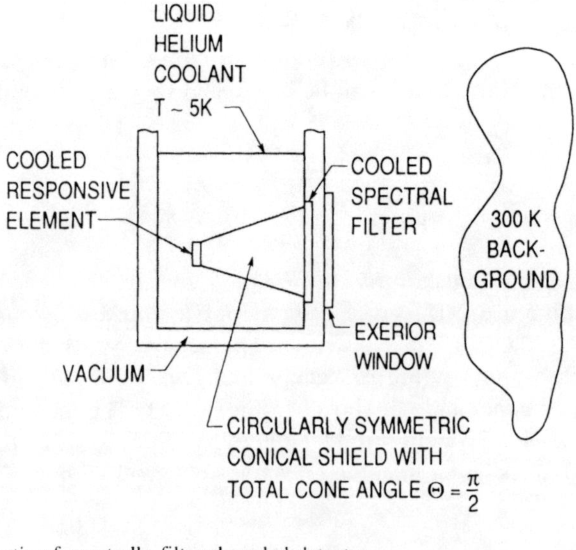

Figure 6.3 Schematic of spectrally filtered cooled detector.

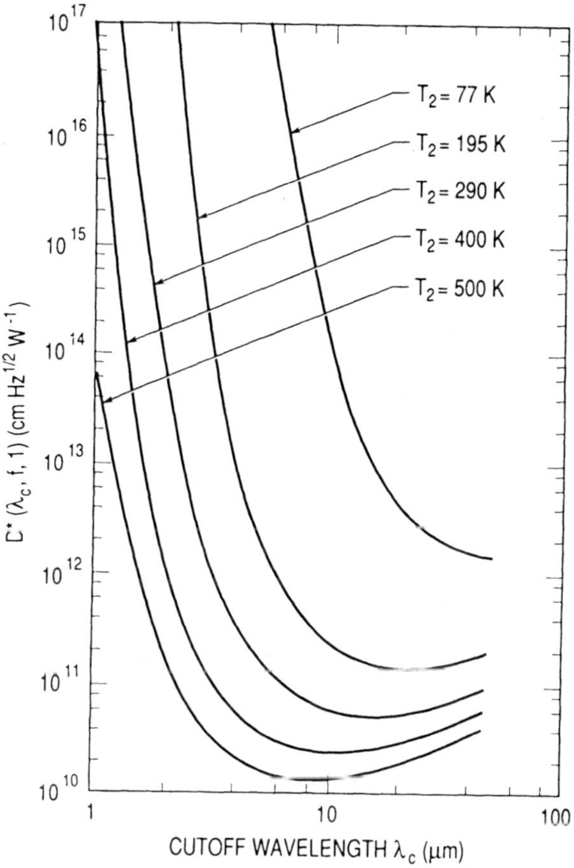

Figure 6.4 Photon noise limited D^* at peak wavelength for various background temperatures.

6.2.2 Example of a S/N Calculation

Consider an electro-optical system used for the detection of point targets having the following system and target parameters:

radiation intensity	= 1 kW
detector instantaneous field of view	= 1 mr × 1 mr
sensor field of view	= 20° az × 5° el
sensor frame time	= 5 s
number of detectors	= 10
optical diamet	= 10 in
focal number ($f/\#$)	= 1
detectivity	= 10^{10} cm-Hz$^{1/2}$/W

$$K_o = 0.5; \quad K_e = 0.7; \quad K_d = 0.84; \quad K_m = 0.84; \quad K_n = 1; \quad K_{fd} = 1 \qquad (6.15)$$

Using these parameters, calculate the S/N ratio at a range of 50 nmi. Assuming that a 90% single-look probability of detection is required for this target at this range, calculate the appropriate threshold setting (in decibels) and also the false alarm rate with this threshold setting.

Substituting eq. (6.14) in (6.13) and using the parameters given in the problem, we have

$$\frac{S}{N} = \frac{\pi(10 \times 2.54)(10^3)(10^{10})(0.5)(0.7)(1)(0.84)(.084)}{4(1)(50 \times 6080/3.28)^2(100)^2} \cdot \{10 \times 5/[(20 \times 5)/(57.3)^2]\} \qquad (6.16)$$

$$= 23.05$$

$$= 13.62 \text{ dB}$$

Conversion factors used in eq. (6.16) include 2.54 cm/in, 6080 ft/nmi, 3.28 ft/m, 100 cm/m, and 57.3 degrees/rad.

From Figure 6.5, a signal-to-noise ratio of 13.6 dB and 90% probability of detection will result in a probability of false alarm of 10^{-7} (point *A* of the figure). The threshold setting for this false alarm probability is given approximately by the 50% probability of detection S/N ratio as 12 dB (point *B* of Figure 6.5).

Using the preceding probability of false alarm, we can compute the *false alarm rate* (FAR) as follows. During each on-target time t_d, the set of detectors n is searched for the existence of a signal. If no signal is present in a detector, the probability of threshold crossing will be due to noise alone, and the FAR will be given as

$$\text{FAR} = \frac{nP_n}{t_d} K_n \qquad (6.17)$$

where

FAR is the false alarm rate (i.e., false alarms per second),
n is the number of detectors,
t_d is the on-target time, s.

Using eq. (6.9) for t_d in (6.17), we get

$$\text{FAR} = \frac{P_n \Omega_s}{\alpha_d T_f} K_n \qquad (6.18)$$

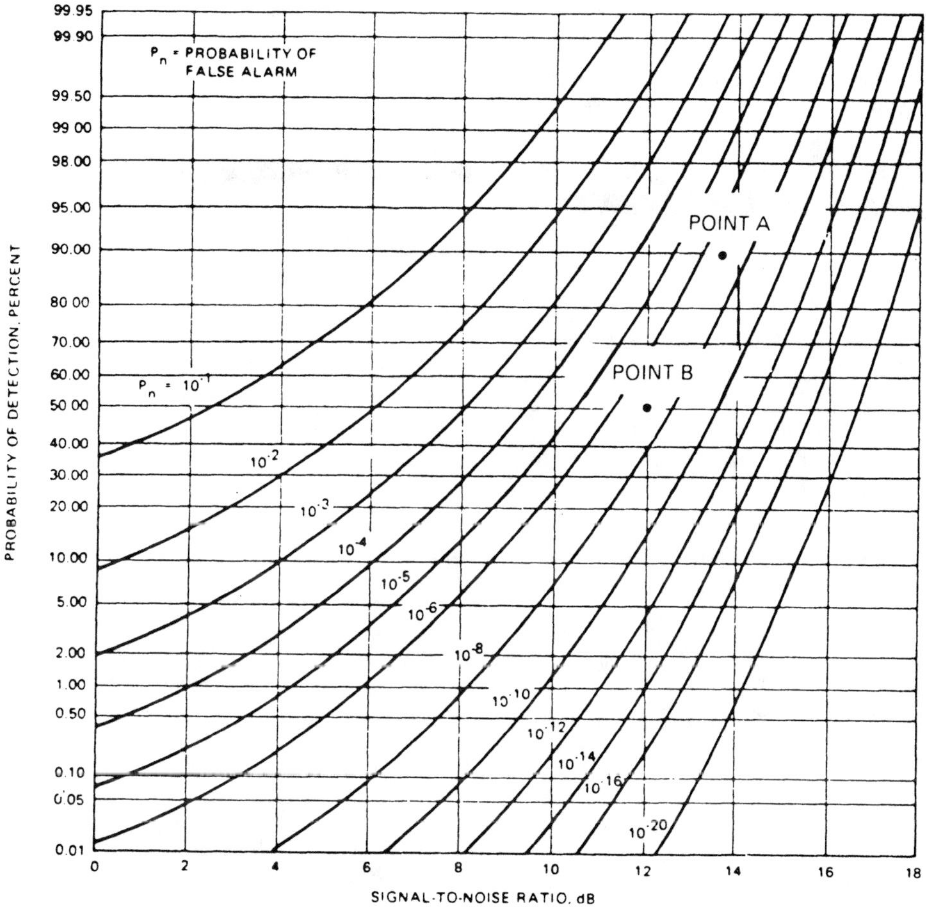

Figure 6.5 Probability of detection versus signal-to-noise ratio.

Using the parameters of the example problem, eq. (6.18), results in

$$\text{FAR} = \frac{10^{-7}\left(\dfrac{20 \times 5}{57.3^2}\right)}{(10^{-3})^2(5)}$$

$$= 6.09 \times 10^{-4} \text{ false alarms per second} \tag{6.19}$$

$$= 2.19 \text{ false alarms per hour}$$

6.2.3 Detection of Satellites from the Ground

As a second example consider a satellite in a 500 km orbit around Earth with a radiation intensity of 500 W/sr. Design an electro-optical sensor in the 8–12 μm range for detection of this satellite.

We start the solution by calculating the range to horizon at which the satellite comes into the view of the sensor. From Figure 6.6, we have

$$R_h = \sqrt{(2\,R_e h + h^2)} \tag{6.20}$$

where

R_h is the range to horizon,
h is the altitude of the satellite,
R_e is the radius of Earth.

(For EO sensor designs requiring orbital calculations, see V. Chobotov, *Orbital Mechanics,* AIAA Education Series, 1991.)

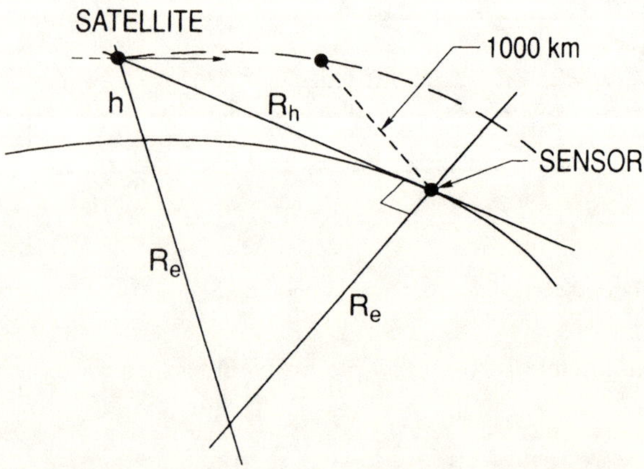

Figure 6.6 Range-to-horizon computation.

Using a height of 500 km and an Earth radius of 6,372 km, we get

$$R_h = 2{,}573 \text{ km} \tag{6.21}$$

Assume that we require detection of satellite on the over pass at a range of 1,000 km and subsequent tracking following this detection. Consider the following design parameters:

satellite radiation intensity	500 W/sr	
detector instantaneous field of view	100×100 μrad	
sensor field of view	$1 \times 1°$	(6.22)
frame time	2 s	
no. of detectors	64	
HgCdTe detector D^*	1.5×10^{10} cm-Hz$^{1/2}$/W	
overall efficiency	0.6	

The HgCdTe detector D^* is obtained from Figure 6.2. We assume that the detector is cooled to 77 K.

The field of view of 1° at a range of 1,000 km is equivalent to a radial distance of 17.45 km. The satellite velocity in km/s can be obtained from the equation:

$$V_s = 7.9\left[1 + \frac{h}{R_e}\right]^{-1/2}$$

At an altitude of 500 km, this results in a satellite velocity of 7.6 km/s. Using this, and the sensor linear field of view, we see that the satellite will be in the field of view for 2 s. This makes the assumed 2 s frame time realistic. Using the parameters of eq. (6.22) in (6.13) we get

$$\frac{S}{N} = \frac{\pi D(500)(1.5E10)(0.6)}{4(f/\#)(1000E5)^2}\left(\frac{64(2)}{\dfrac{1 \times 1}{(57.3)^2}}\right)^{1/2} \tag{6.23}$$

$$= 0.229\frac{D}{f/\#}$$

Using a focal number of 2 and solving for D we get

$$D = (S/N)/0.1145 \tag{6.24}$$

We need to detect this satellite within the first frame of 2 s with a high probability of detection; that is 90%. From Figure 6.5 we may select

$$P_d = 90\%; \quad S/N = 13.5 \text{ dB}; \quad P_n = 10^{-7} \tag{6.25}$$

where P_d is the probability of detection, and P_n is the probability of false alarm. The threshold setting corresponding to a 50% probability of detection for a false alarm probability of 10^{-7} will be 12 dB.

Using eq. (6.24) with a S/N of 13.5 dB (22.38) we get a lens diameter of

$$D = 195.5 \text{ cm}$$

$$= 2 \text{ m} \tag{6.26}$$

The diffraction limit of this lens, or the null-to-null beamwidth γ_{nn}, as it sometimes is referred to, is

$$\gamma_{nn} = \frac{k\lambda}{D} \tag{6.27}$$

where λ is the wavelength, and D is the lens diameter. The value of k is 2 for a $\sin x/x$ beam pattern and may be higher for a Gaussian or other pattern distributions. In optical system it is usually taken to be

$$\gamma_{nn} = \frac{2.44\lambda}{D} \tag{6.28}$$

Using this for the wavelength center of 10 μm, which is between 8 and 12 μm, we get

$$\gamma_{nn} = 12.2 \ \mu\text{rad} \tag{6.29}$$

which is considerably less that the instantaneous field of view of the detector, which was 100 μrad. Therefore, the target energy through the lens will be within the detector FOV.

The false alarm rate can be computed from eq. (6.18) as follows:

$$\text{FAR} = \frac{(10^{-7})(1 \times 1)}{(57.3)^2(100E - 6)^2(2)}$$

$$= 1.3E - 3 \text{ alarms per second} \tag{6.30}$$

$$= 5 \text{ alarms per hour}$$

Assuming this value is acceptable, the design can be considered complete.

Variations on this design can also be considered. For example, assume that we use a 1 s scan time and require detection on two consecutive scans. To get 90% detection on two consecutive frames, we need a probability of detection that is the square root of this in each frame, or a P_d of 95%. The corresponding false alarm rate can also be reduced to square root of 10^{-7} or 3.16×10^{-4}. From Figure 6.5 for a P_d of 95% and a P_n of 3.16×10^{-4} we need a S/N of 12 dB. The threshold setting will be about 10 dB. The frame time of 1 s will result in a corresponding equation to (6.24) as

$$\mathrm{D} = (S/N)/0.081 \tag{6.31}$$

Using S/N of 12 dB (15.84) in eq. (6.31) we get

$$D = 195.55 \text{ cm}$$

$$= 2 \text{ m} \tag{6.32}$$

Note that this design resulted in the same lens size and the same false alarm rate. The only difference is the lower frame time and the additional logic for two out of two detections. This design may be preferable, as it will also detect the target overhead where the linear field of view is (500) (1/57.3) = 8.72 km and the satellite stays in the field of view for slightly over 1 s.

6.3 NOISE EQUIVALENT FLUX DENSITY

Another often-used system performance measure is the noise equivalent flux density. This value is defined as

$$\mathrm{NEFD} = \frac{H_t}{S/N} \tag{6.33}$$

Using eqs. (6.1) and (6.13) in (6.33), we get

$$\mathrm{NEFD} = \frac{4(f/\#)}{\pi D D^* L}\left(\frac{\Omega_s}{nT_f}\right)^{1/2} \tag{6.34}$$

The S/N equation and the NEFD equation are used to predict the performance of existing electro-optical systems or to design new systems.

6.4 SPECTRAL EMITTANCE OF A TARGET

The spectral emittance of a thermal target can be written

$$W(\lambda, T) = \varepsilon(\lambda) W_b(\lambda, T) \tag{6.35}$$

where

$W(\lambda, T)$	is the target spectral emittance, W/cm^2-sr-μm,
$\varepsilon(\lambda)$	is the target emissivity,
$W_b(\lambda, T)$	is the blackbody spectral emittance, W/cm^2-sr-μm,
T	is the target temperature, K,
λ	is the wavelength, μm.

The spectral emittance $W(\lambda, T)$ is usually used for distributed thermal targets, and it is obtained by measurements as shown in Figure 6.7 [2]. This figure gives the spectral emittance of a grass-covered field at two different temperatures, 32°C

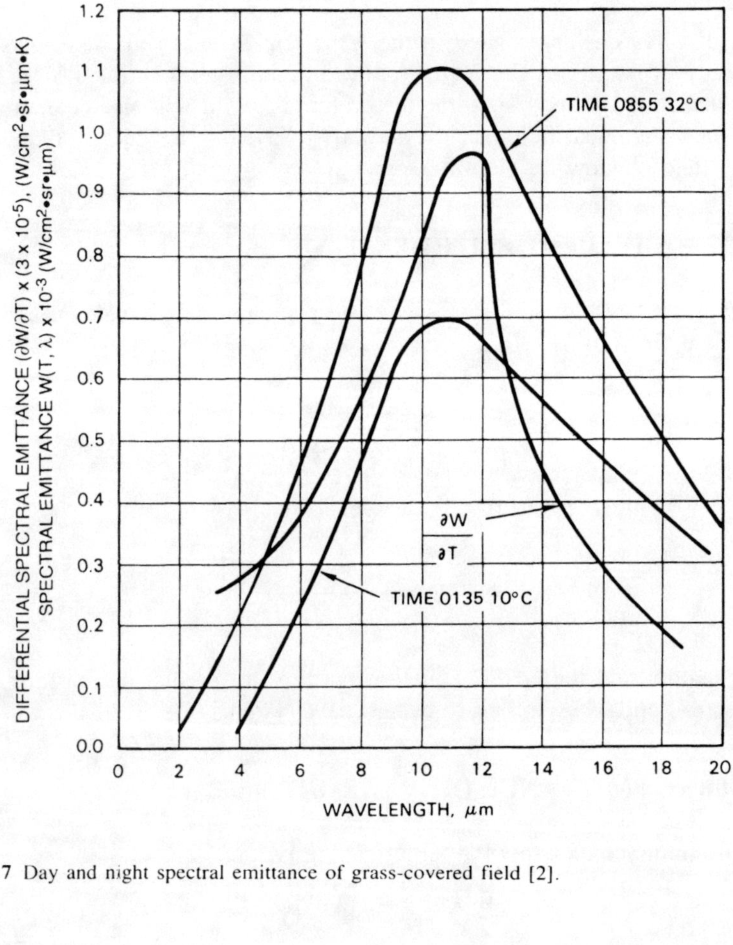

Figure 6.7 Day and night spectral emittance of grass-covered field [2].

and 10°C. The spectral emittance can be applied in the calculation of the target signal power as follows:

$$S = \frac{\pi D^2}{4}\alpha_d \int_{\lambda_1}^{\lambda_2} W(\lambda, T)\, \mathrm{d}\lambda \tag{6.36}$$

where D is the aperture diameter and α_d, as before, is the instantaneous field of view. The units of the signal S of eq. (6.36) will be in Watts. The integral is taken over the range of detector wavelengths.

Using eq. (6.36), together with the noise power equation (6.12), we can obtain the signal-to-noise ratio for a distributed target:

$$\frac{S}{N} = \frac{\pi D \alpha_d L \tau_a}{4(f/\#)}\left(\frac{nTf}{\Omega_s}\right)^{1/2} \int_{\lambda_1}^{\lambda_2} W(\lambda, T) D^*(\lambda)\, \mathrm{d}\lambda \tag{6.37}$$

The coefficient τ_a, denoting the optical transmission efficiency, is introduced in eq. (6.37).

Using the spectral emittance curves of Figure 6.7, the differential spectral emittance $\partial W(\lambda)/\partial T$ can be calculated and plotted as shown. This curve is used in another system performance measure, called *noise equivalent temperature* (NET) as will be seen shortly.

The incremental target emittance can be obtained by using the differential of equation (6.35) for a given wavelength, λ:

$$\mathrm{d}W(\lambda, T) = \frac{\partial W(\lambda, T)}{\partial \varepsilon}\partial \varepsilon + \frac{\partial W(\lambda, T)}{\partial T}\mathrm{d}T \tag{6.38}$$

Using the difference notation, this becomes

$$\Delta W(\lambda, T) = \frac{\partial W(\lambda, T)}{\partial \varepsilon}\Delta \varepsilon + \frac{\partial W(\lambda, T)}{\partial T}\Delta T \tag{6.39}$$

Assuming that the background emissivity remains constant ($\Delta\varepsilon = 0$), the spectral emittance will change only with temperature, as follows:

$$\Delta W(\lambda, T) = \frac{\partial W(\lambda, T)}{\partial T}\Delta T \tag{6.40}$$

Using this, the change in signal power due to an incremental temperature change

ΔT can be computed by using the signal power equation:

$$\Delta S = \frac{\pi D^2}{4} \alpha_d \int_{\lambda_1}^{\lambda_2} \frac{\partial W(\lambda, T)}{\partial T} \Delta T \, d\lambda \tag{6.41}$$

where ΔS is the incremental signal change due to temperature change of ΔT. Equation (6.41) can also be obtained directly by differentiating the signal power equation (6.36) with respect to T.

Using the noise power equation (6.12), the target signal power equation (6.41), and the optical transmission efficiency, τ_a, we obtain

$$\frac{\Delta S}{N} = \frac{\pi D \alpha_d \tau_a D^* L}{4(f/\#)} \left(\frac{n T_f}{\Omega_s} \right)^{1/2} \int_{\lambda_1}^{\lambda_2} \frac{\partial W}{\partial T} \Delta T \, d\lambda \tag{6.42}$$

where the loss term L is incorporated in the numerator of the equation.

6.5 NOISE EQUIVALENT TEMPERATURE

Another measure of optical system performance is depicted by the noise equivalent temperature and can be derived easily using the $\Delta S/N$ equation (6.42). Assuming the detectivity D^* to be a function of wavelength, eq. (6.42) can be solved for ΔT as a function of $\Delta S/N$. The value of ΔT thus obtained for a $\Delta S/N$ of unity is termed the *noise equivalent temperature*, or

$$\text{NET} = \frac{4(f/\#)}{\pi \tau_a D \alpha_d L} \left(\frac{\Omega_s}{n T_f} \right)^{1/2} \frac{1}{\displaystyle\int_{\lambda_1}^{\lambda_2} \frac{\partial W(\lambda, T)}{\partial T} D^*(\lambda) \, d\lambda} \tag{6.43}$$

Equation (6.43) can also be used as a design tool to establish requirements of electro-optical systems as shown in the following example.

Spacecraft Imaging System Example

Consider the design of an IR imaging system on board a spacecraft at 500 mi altitude and having a velocity of 4 miles per second [4]. The azimuth coverage is 12° and the elevation coverage is that obtained for 1 s of space flight. We want to obtain a

NET of 0.10 K. Consider the following design parameters obtained after several iterations:

$$
\begin{aligned}
&\text{diameter of the optics} = 8 \text{ in} \\
&\text{number of detectors} = 50 \\
&\text{spatial resolution} = 200 \text{ ft} \\
&f/\# = 1 \\
&\text{number of frames} \\
&\text{per second} = 30 \\
&\text{peak detectivity} = 2 \times 10^{11} \text{ cm-Hz}^{1/2}/\text{W}
\end{aligned}
\tag{6.44}
$$

$\tau_a = 0.7$; $K_o = 0.6$; $K_d = 0.8$; $K_m = 0.8$; $K_e = 0.8$; $K_n = 0.81$; $K_{fd} = 1.5$

Additionally, the normalized differential emittance and detector detectivity curves of Figure 6.8 are also given. Assume detectors are operating in the wavelength region of 8 to 14 μm.

Using the preceding data, we can calculate a number of additional parameters to be used in conjunction with NET equation (6.43). From spacecraft velocity and altitude, we can calculate the equivalent elevation coverage for 1 s of flight:

$$
\begin{aligned}
\varepsilon &= \frac{Vt}{h} \\
&= \frac{4(1)}{500} = 8 \times 10^{-3} \text{ rad}
\end{aligned}
\tag{6.45}
$$

Using this and the azimuth coverage of 12°, we obtain the spatial window:

$$\Omega_s = \frac{12}{57.3}(8 \times 10^{-3}) = 1.675 \times 10^{-3} \text{ sr} \tag{6.46}$$

The required resolution of 200 ft, will give the instantaneous detector field of view:

$$\alpha_d^{1/2} = \frac{200}{(500)(5280)} = 7.575 \times 10^{-5} \text{ radians} \tag{6.47}$$

where 5,280 is conversion from miles to feet. Using eq. (6.47), α_d becomes

$$\alpha_d = 5.74 \times 10^{-9} \text{ sr} \tag{6.48}$$

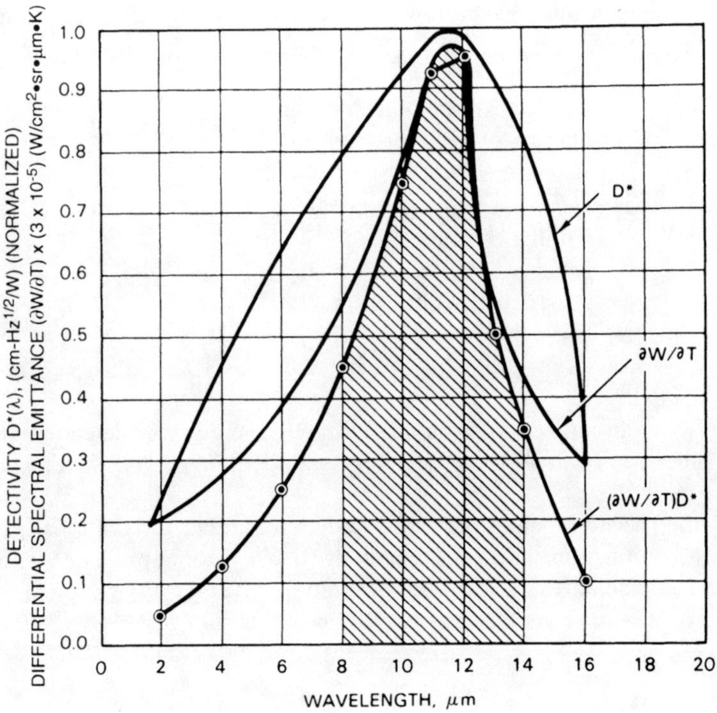

Figure 6.8 Spectral emittance and detectivity versus wavelength.

Using the preceding values, together with $L = 0.225$ computed from eq. (6.14), in eq. (6.43), we can write

$$\text{NET} = \frac{4(1)}{\pi(0.7)(8 \times 2.54)(5.74\,E - 9)(0.225)} \left(\frac{1.675 \times 10^{-3}}{50(1/30)} \right)^{1/2} \times \left(\frac{1}{(2 \times 10^{11})(1.2 \times 10^{-4})} \right) \tag{6.49}$$

$$= 0.0916\ \text{K}$$

In this equation, 1.2×10^{-4} represents the value of the integral in the denominator of eq. (6.43) obtained by calculating the area under the curve of Figure 6.8 (dashed region). The multiplier of this value is the normalization constant $D*$ (λ) given in eq. (6.44).

The NET value of eq. (6.49) is reasonably close to the design goal of 0.10 K.

From the given parameters, a number of other system constants can also be derived. For example, the detector on-target time from eq. (6.9) becomes

$$t_d = \frac{n\alpha_d T_f}{\Omega_s}$$

$$= \frac{(50)(5.74 \times 10^{-9})(1/30)}{1.67 \times 10^{-3}} \tag{6.50}$$

$$= 5.711 \times 10^{-6} \text{ s}$$

Using this value in eq. (6.10) with noise bandwidth factor of 1.5, we get

$$\Delta f = \frac{1.5}{t_d}$$

$$= 262 \text{ kHz} \tag{6.51}$$

representing the bandwidth of the received signal.

6.6 MINIMUM RESOLVABLE TEMPERATURE

The minimum resolvable temperature can be derived from the noise equivalent temperature by considering the display and observer parameters.

Assuming a target resolution distance T_x from an observation point at a distance of R (see Figure 6.9), we get

$$\theta_d^{1/2} = \frac{1000T_x}{R} \tag{6.52}$$

where

$\theta_d^{1/2}$ is the angular resolution, mrad,
$T\chi$ is the linear resolution distance at the target, m,
R is the distance to the observation point, m.

The one-half power notation is used in eq. (6.52) to conform with the previous instantaneous field-of-view notation of $\alpha^{1/2}$. The one-half power designation refers

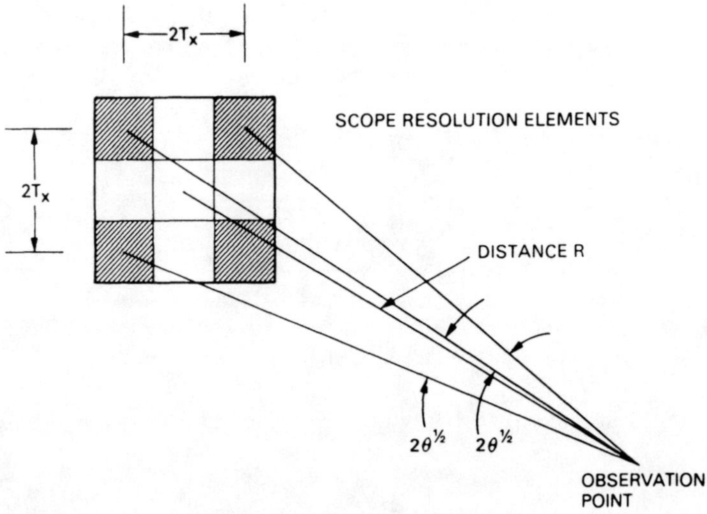

Figure 6.9 Target resolution elements and subtended angles.

to a linear angular measurement, whereas the unity power refers to a two-dimensional measurement. Using the angular resolution of eq. (6.52), we can define a spatial frequency as

$$f_s = \frac{1}{2\theta_d^{1/2}} \tag{6.53}$$

where f_s is the spatial frequency, cycles/mrad.

For a square display cell, the subtended angle will be $\theta_d^{1/2}$ mrad. This spatial frequency through eqs. (6.52) and (6.53) is related to the thermal image-target distance R. Because the NET of the target is calculated per detector resolution cell, the equivalent display NET (referred to as *MRT*) will be modified by $\theta_d^{1/2}$ as follows:

$$\text{MRT} = (\text{NET})(\alpha_d/\theta_d)^{1/2} \tag{6.54}$$

where

$\alpha_d^{1/2}$ is the detector field of view,
$\theta_d^{1/2}$ is the display resolution cell.

Another factor affecting the MRT is the observer integration of the detected signal. Noting that the NET value was calculated for a single detector frame, this value should be modified by the number of frames integrated by the observer. The number of integrated frames can be written

$$n_i = T_e\dot{F} \tag{6.55}$$

where

n_i is the number of frames integrated by the observer,
T_e is the observer integration time, s,
$\dot{F}$ is the frame rate, frames per second.

Using eq. (6.55) in (6.54), we get

$$\mathrm{MRT} = (\mathrm{NET})\left(\frac{\alpha_d}{\theta_d}\right)^{1/2}\left(\frac{1}{T_e\dot{F}}\right) \tag{6.56}$$

From detection studies performed by a number of researchers [5–7] on the detection and recognition of targets in a given scene, such as a TV picture, it was concluded that a S/N ratio of 2 or 3 is needed to detect or recognize an object from its background. The background may be a field, a road, or sand on which the target is located. From Figure 6.5 a value of S/N ratio of 3 (4.8 dB) will correspond to about 50% probability of detection with a false alarm probability of 5×10^{-2}. Using this value it is assumed that, on the average, about 50% of resolution cells covering the target would be sufficient for detection. Incorporating this factor in eq. (6.56) and using spatial frequency relation (6.53), we get

$$\mathrm{MRT} = \frac{6(\mathrm{NET})f_s(\alpha_d)^{1/2}}{(T_e\dot{F})^{1/2}} \tag{6.57}$$

Replacing the NET of this equation by its equivalent expression from eq. (6.43), and using (6.14) for L, we get

$$\mathrm{MRT} = \frac{24(f/\#)}{\pi\tau_a D}\left(\frac{f_s}{r_s}\right)\left(\frac{K_{fd}}{K_n}\right)^{1/2}\left(\frac{\Omega_s}{nT_f\alpha_d}\right)^{1/2}\left(\frac{1}{T_e\dot{F}}\right)^{1/2}\frac{1}{\displaystyle\int_{\lambda_1}^{\lambda_2}\frac{\partial W(\lambda, T)}{\partial T}D^*(\lambda)\,d\lambda} \tag{6.58}$$

This expression for MRT includes optical, detector, electronics, and monitor losses as these values are included in the r_s (modulation transfer function MTF) term. The overall system MTF, as discussed at length in previous chapters, can be computed as a function of spatial frequency of eq. (6.53).

An example of MTF curve is given in [5] and reproduced here as Figure 6.10. The MTF curve of Figure 6.10 is normalized with respect to the limiting spatial frequency f_{ls} defined as follows:

$$\bar{f}_s = f_s/f_{ls} \qquad (6.59)$$

where $\bar{f}_s$ is the normalized spatial frequency, f_s is the spatial frequency, and f_{ls} is the limiting spatial frequency. The value of f_{ls} can be defined in terms of optical aperture (for optics limited designs) and the wavelength as

$$f_{ls} = \frac{D}{\lambda} \qquad (6.60)$$

where f_{ls} is the limiting spatial frequency, D is the optical aperture, and λ is the wavelength. The limiting spatial frequency f_{ls} should be obtained from the detector IFOV for detector limited designs.

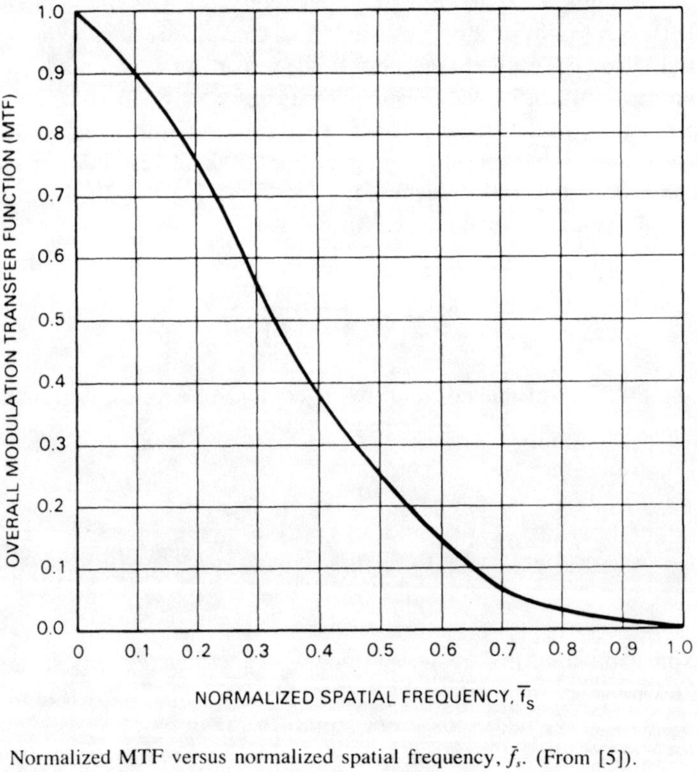

Figure 6.10 Normalized MTF versus normalized spatial frequency, $\bar{f}_s$. (From [5]).

The value of MRT obtained from eq. (6.58) is another important parameter used in performance and design analysis of electro-optical systems.

6.7 IMAGE QUALITY CONSIDERATIONS

An important consideration in image quality analysis is target resolution. Figure 6.11 illustrates the problem of resolving two closely spaced targets with unequal signal intensity. The optics spreads the images of the two objects at the focal plane so that they overlap when in close proximity. The focal plane images are sampled by detectors of finite size. A mosaic detector array is shown but a single detector can also be used to sample the images.

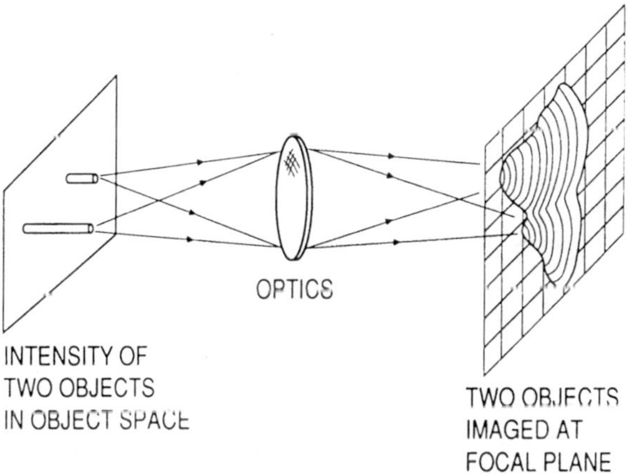

Figure 6.11 Resolution problem of two closely spaced targets of unequal intensity.

The general factors that must be considered are

1. The optics point spread function typically includes both the ideal diffraction and geometric distortions due to imperfect optics. The latter can cause changes in the point spread images as a function of position in the image field. In general, the focal surface may not be planar.
2. *Line of sight* (LOS) jitter can cause the images to bounce around.
3. The finite detector size can limit the sampling precision of the images.
4. System noise consists of Gaussian signal noise, detector noise, and noise from other sources that are system dependent, such as power supply and digital switching mechanism.
5. Sensor operating modes and characteristics include scanning sensor or staring sensor, frame rate, sample rate, digitizer quantization, and dynamic range.

In general, for a square detector, the resolution performance of a scanning sensor is better in the scan direction (target horizontal) and poorer in the cross-scan direction (target vertical) as illustrated in Figure 6.12.

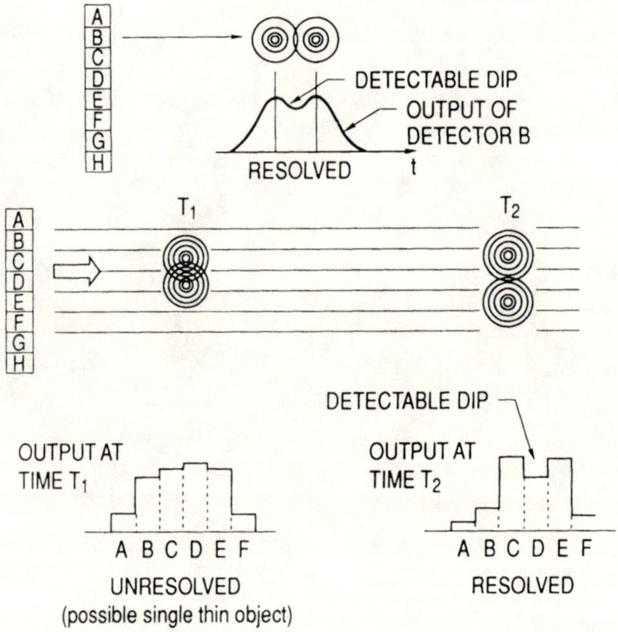

Figure 6.12 Scanning sensor resolution for objects aligned in scan and cross-scan directions.

As shown, the phasing or relative vertical position of the detectors to targets is critical in determining whether there is one thin object or two separate point sources.

The preceding example also introduces another uncertainty factor: the physical extent of the object. If it is known a priori that all objects are point sources, then it is possible to determine whether one or more objects are present. By comparing the known responses of a single point source object, such as its 50% amplitude width, multiple object images can be distinguished from that of the single object.

The resolution of targets, as discussed earlier, is only one consideration of image quality. As is well known, most targets are not point sources, especially in imaging sensors. Point source targets primarily occur when the detector instantaneous field of view is considerably larger than the target size. This is the case in *infrared search and tracking systems* (IRST). In imaging sensors of the electro-optical type, usually the image of the scene with the object is produced. From this scene we need to identify the targets of interest. In general several categories of detections, orientation, recognition and identification can be defined as follows:

- *Detection*—the discovery of the existence of an object without complete recognition of the object: "Something is there."
- *Orientation*—orientation of the object, such as the side or front view.
- *Recognition*—determination by any means of the type of the target such as cars, trucks, tanks, aircraft, and so on.
- *Identification*—the discrimination between recognized types of targets, such as a US aircraft or a French aircraft.

Each of these categories requires a different level of target resolution, and they become progressively more difficult as one moves from simple detection to identification. The main categories of image quality consideration are detection and recognition.

Figure 6.13 gives an illustration of the number resolution cells (pixels) needed for detection, orientation, recognition, and identification. We see that advancing from one category to the next requires additional pixels on the target.

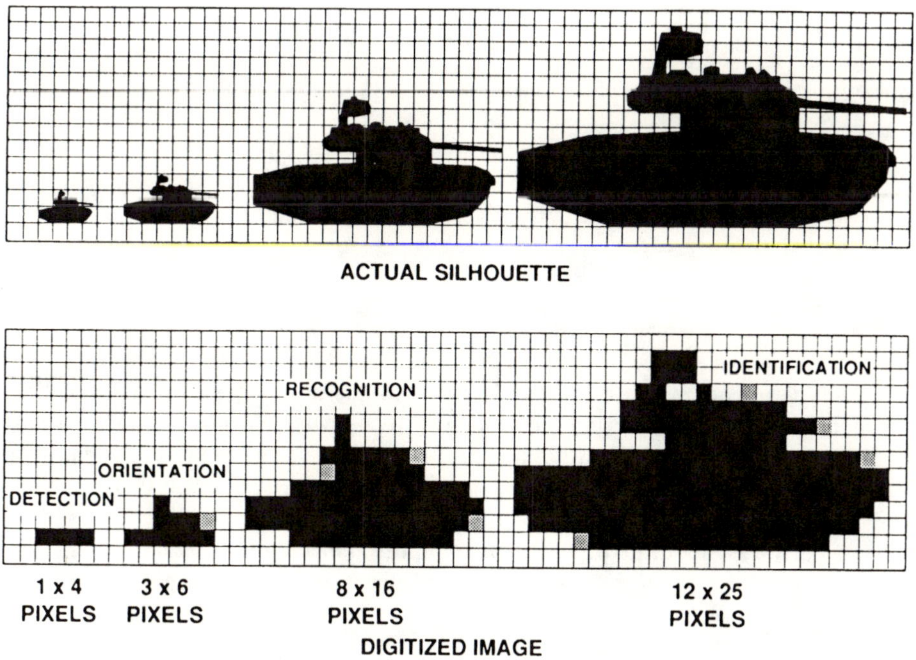

Figure 6.13 Object size relative to resolution element.

Usually the number of pixels required per minimum target dimension is specified in image quality considerations. In a number of cases, however, another dimension is defined as the critical dimension, replacing the minimum target dimension value. This is especially true for airborne and spaceborne sensors, where the "look angle" of the sensor sees the target from the top. For example consider an automobile with dimensions of

height	1.25 m
width	2 m
length	6 m

The minimum dimension will be 1.25 m. But, a logical frontal critical dimension may be

$$C_f = \sqrt{(2^2 + 1.25^2)}$$

$$= 2.35 \text{ m}$$

Perhaps a logical critical dimension from an airborne or spaceborne sensor may be

$$C_{as} = \sqrt{(2^2 + 6^2)}$$

$$= 6.32 \text{ m}$$

The number of pixels, line pairs, or cycles required for detection, recognition, and identification of targets is usually obtained experimentally as follows.

6.8 EXPERIMENTAL RESULTS

Through experiments with observers, Johnson [8] determined the number of line pairs required to detect, determine orientation, recognize, and identify tactical targets. These results are illustrated in Figure 6.14. We can see from the accompanying table that the requirements in each category are relatively independent of the target. Recognition, which is the most frequently used figure of merit for system performance calculations, requires four line pairs across the minimum target dimension. On the other hand, simple target detection requires only one line pair across the minimum target dimension. The table of Figure 6.14 also gives the average values of line pairs (cycles) across the minimum target dimension for detection, orientation, recognition, and identification. Figure 6.15 is a plot of the values of Figure 6.14. From Figure 6.15 we see that the variations about the mean values increase as categories of target description improves from detection to identification.

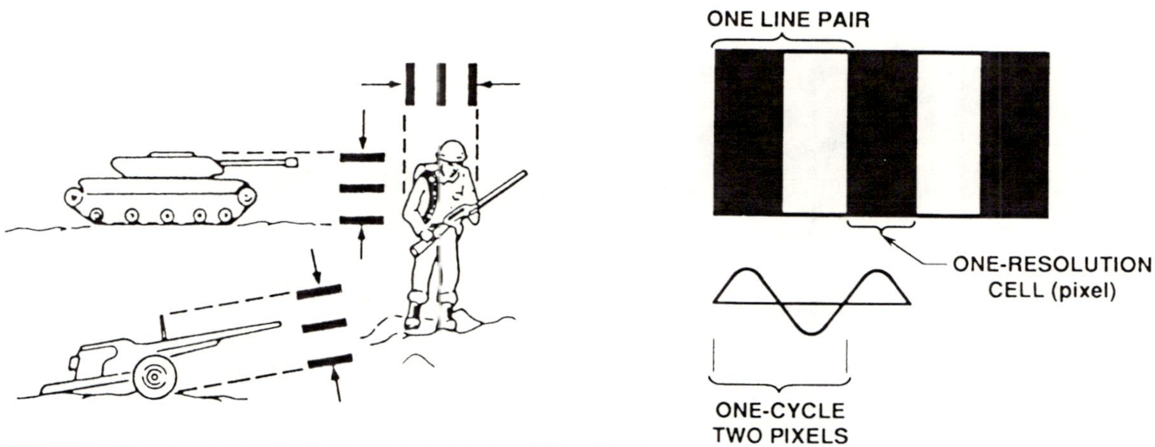

TARGET	RESOLUTION PER MINIMUM DIMENSION IN LINE PAIRS			
BROADSIDE VIEW	DETECTION	ORIENTATION	RECOGNITION	IDENTIFICATION
TRUCK	0.90	1.25	4.5	8.0
M-48 TANK	0.75	1.20	3.5	7.0
STALIN TANK	0.75	1.20	3.3	6.0
CENTURION TANK	0.75	1.20	3.5	6.0
HALF-TRACK	1.00	1.50	4.0	5.0
JEEP	1.20	1.50	4.5	5.5
COMMAND CAR	1.20	1.50	4.3	5.5
SOLDIER (standing)	1.50	1.80	3.8	8.0
105 HOWITZER	1.00	1.50	4.3	6.0
AVERAGE	1.0 ± 0.25	1.4 ± 0.35	4.0 ± 0.8	6.4 ± 1.5

Figure 6.14 Required resolution for detection, orientation, recognition, and identification.

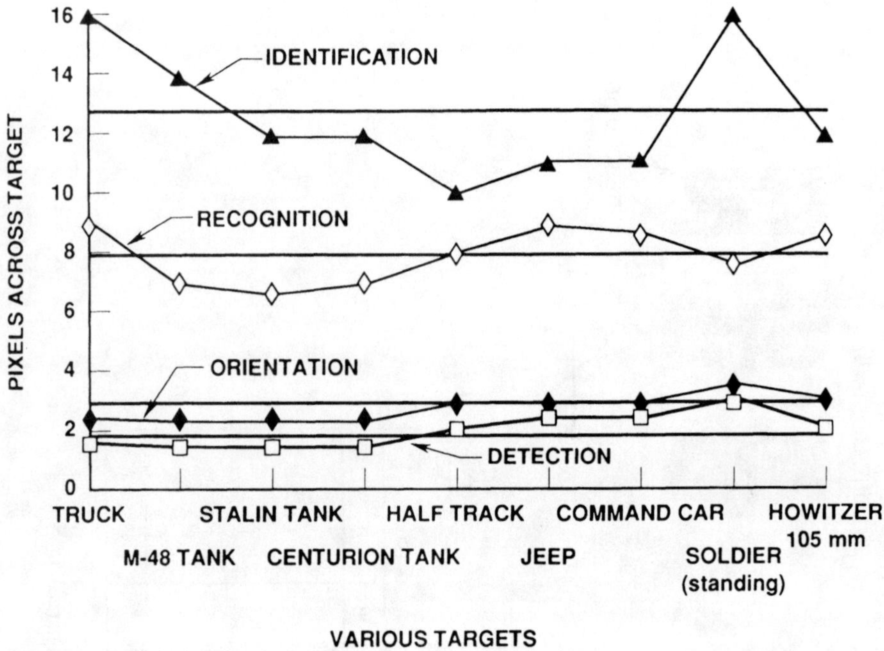

Figure 6.15 Johnson's criterion for tactical targets.

6.9 NUMERICAL EXAMPLE OF MRT CALCULATION

As a numerical example of MRT calculations, consider a thermal imaging system with the following parameters:

$$
\begin{aligned}
&\text{spectral band} && = 8\text{–}12\ \mu\text{m} \\
&\text{optical diameter } (D) && = 8 \text{ in} \\
&\text{focal number } (f/\#) && = 2 \\
&\text{filter bandwidth factor } K_{fd}) && = 1 \\
&\text{scanning efficiency } (K_n) && = 0.8 \\
&\text{sensor field of view} && = 2.0^\circ \text{ (az)},\ 1.0^\circ \text{ (el)} \\
&\text{number of detectors } (n) && = 1{,}024 \\
&\text{frame time } (T_f) && = 3.34 \times 10^{-2} \text{ s} \\
&\text{detector instantaneous field of view } (\alpha_d^{1/2}) && = 50\ \mu\text{rad} \\
&\text{frame rate } (\dot{F}) && = 30 \text{ per second} \\
&\text{eye integration time } (T_e) && = 0.2 \text{ s} \\
&\text{peak detectivity } (D^*) && = 2 \times 10^{11} \text{ cm-Hz}^{1/2}/\text{W} \\
&\text{optical efficiency } (\tau_a) && = 0.5
\end{aligned}
\tag{6.61}
$$

In addition to these parameters, use the spectral emittance and detectivity characteristics of Figure 6.8. Using this figure,

$$\int_{8}^{12} \frac{\partial W[\lambda, T]}{\partial T} D^*(\lambda)\, d\lambda = 8.4 \times 10^{-5}\ \text{Hz}^{1/2}/\text{cm-sr-K} \tag{6.62}$$

Note that the integral of eq. (6.62) is taken between the wavelength limits of 8 and 12 μm rather than 8 and 14 μm shown by the dashed area of Figure 6.8.

Use an atmospheric transmission coefficient of unity and calculate the atmospheric losses as discussed in the following section.

Using the preceding parameters, calculate the detection and recognition range of the system for a tank-type target with a critical dimension of 3 m with a target to background temperature difference ΔT of 2.5°.

Using the system parameters of eq. (6.61), we can calculate the parameters of MRT equation (6.58); that is,

$$\begin{aligned} \Omega &= \frac{2 \times 1}{(57.3)^2} \\ &= 6.09 \times 10^{-4}\ \text{sr} \\ \alpha_d &= (50 \times 10^{-6})^2 \\ &= 2.5 \times 10^{-9}\ \text{sr} \end{aligned} \tag{6.63}$$

Using the parameters given in eq. (6.58), we get

$$\text{MRT} = \frac{(24)(2)}{\pi(8 \times 2.54)} \left(\frac{f_s}{r_s}\right) \left(\frac{1}{0.8}\right)^{1/2} \left(\frac{6.09 \times 10^{-4}}{(1024)(3.34 \times 10^{-2})(2.5 \times 10^{-9})}\right)^{1/2}$$

$$\left(\frac{1}{0.2 \times 30}\right)^{1/2} \left(\frac{1}{(8.4 \times 10^{-5})(2 \times 10^{11})}\right)$$

$$= (1.724E - 6)\left(\frac{f_s}{r_s}\right) (f_s \text{ in cycles/rad}) \tag{6.64}$$

$$= (1.724E - 3)\left(\frac{f_s}{r_s}\right) (f_s \text{ in cycles/mrad})$$

Using eq. (6.64), with a limiting spatial frequency of

$$f_{ls} = \frac{D}{\lambda}$$

$$= \frac{8 \times 2.54 \times 10^{-2}}{10 \times 10^{-6}} \tag{6.65}$$

$$= 20.32 \text{ cycles/mrad}$$

together with r_s values of Figure 6.10, we get the spatial frequency f_s and the corresponding MRT values of Table 6.1. These values are plotted on Figure 6.16, showing the exponential nature of the MRT curve.

Table 6.1
Spatial Frequency and MRT

Spatial Frequency, f_s	*MRT K*	*Spatial Frequency, f_s*	*MRT K*
1	0.00182	11	0.08225
2	0.00384	12	0.12918
3	0.00618	13	0.22421
4	0.00909	14	0.40287
5	0.01287	15	0.64810
6	0.01812	16	0.92253
7	0.02335	17	1.47138
8	0.03253	18	3.11792
9	0.04404	19	—
10	0.05919	20	—

Noting that the target has a dimension of 3 m, each spatial frequency cycle will represent a range of 3 km. The reason is that each cycle represents a 1-mrad subtended angle.

From Figure 6.16, the given tank ΔT of 2.5° will correspond to a spatial frequency of 17.6 cycles/mrad. From Figure 6.14, on the average, one line pair or one cycle is required for *detection*. Using this, the detection range of the thermal sensing system becomes

$$(3 \text{ km})\,(17.6 \text{ cycles}) = 52.8 \text{ km} \tag{6.66}$$

where 3 km represents the corresponding range for the critical dimension of 3 meters.

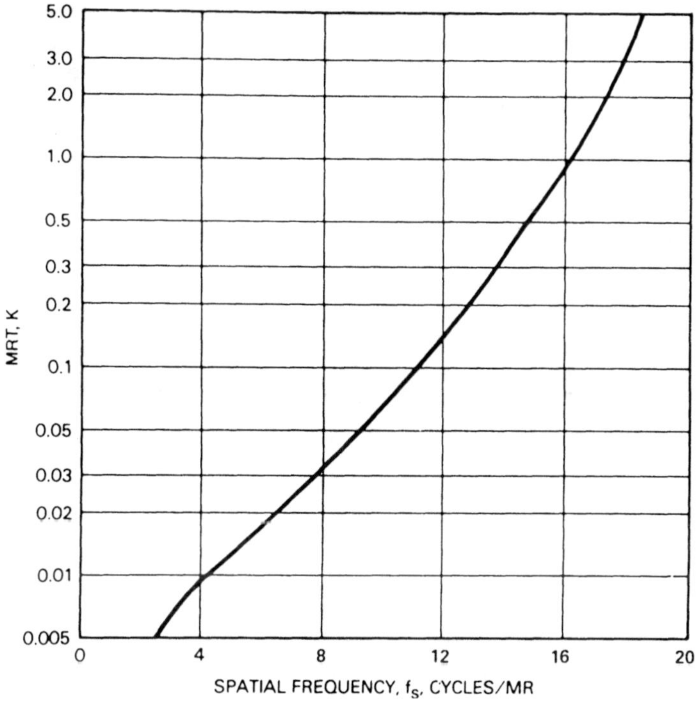

Figure 6.16 MRT versus spatial frequency for numerical example.

From Figure 6.14, the average number of *recognition* line pairs is four. Using this, the recognition range becomes

$$(3)\,(17.6/4) = 13.2 \text{ km} \tag{6.67}$$

Note that the detection and recognition values given in eqs. (6.66) and (6.67) do not include atmospheric losses, which will considerably reduce these ranges as will be seen in the next section.

Atmospheric Loss Calculation

Atmospheric loss factor considerations were discussed in Chapter 3. From Table 3.4 consider the case of visibility range of greater than 4 km, the air temperature greater than 10° C and relative humidity of 50–80%. In this case the corresponding atmospheric attenuation loss will be 0.9 dB/km.

The trial-and-error calculations resulting in the detection ranges, which includes atmospheric losses, are given in Table 6.2. This table includes target sensor range and the corresponding atmospheric transmission loss calculated by using a 0.9 dB/km loss factor. The value of $\bar{\Delta T}$, which includes the transmission loss, is obtained by multiplying the design ΔT of 2.5° by the transmission loss factor. Using this $\bar{\Delta T}$ value and the curve of Figure 6.16 (or a computer printout), the new value of spatial frequency is obtained. Multiplying this value by 3 km will result in the detection range with atmospheric losses.

Table 6.2

Calculation of Atmospheric Losses and Corresponding Detection Range

Target-Sensor Range, km	*Transmission Loss Factor,* τ_a	$\bar{\Delta T} = \Delta T \times (\tau_a)$	*Spatial Frequency*	*New Range, km*
10	0.126	0.315	13.6	40.8
11	0.102	0.255	13.2	39.6
12	0.083	0.207	12.8	38.4
13	0.068	0.170	12.4	37.2
14	0.054	0.135	12.0	36.0
15	0.045	0.112	11.8	35.4
16	0.036	0.090	11.2	33.6
17	0.029	0.072	10.5	31.5
18	0.024	0.060	9.95	29.85
19	0.019	0.047	9.14	27.42
20	0.016	0.040	8.53	25.59
21	0.013	0.032	7.92	23.76
22	0.010	0.025	7.10	21.30
23	0.008	0.020	6.50	19.50
24	0.007	0.017	5.90	17.70
25	0.006	0.015	5.50	16.50

From Table 6.1, we see that a possible match exists between starting and final target sensor range of about 21 km. Thus, the detection range of the thermal sensor under the stated atmospheric condition is approximately 21 km as opposed to the no-loss range of 52.8 km.

In a similar manner, recognition ranges, which include atmospheric losses, can be calculated as shown in Table 6.3. Because the no-loss recognition range from eq. (6.67) was about 13 km, the target-sensor ranges of Table 6.3 were taken around this value. The atmospheric transmission losses were for a 0.9 dB/km loss factor, and a four-cycle recognition criterion was used in computing the new range values of this table. From Table 6.3, we see that a match is obtained at a range of about

Table 6.3
Calculation of Atmospheric Losses and the Corresponding Recognition Range

Target-Sensor Range, km	*Transmission Loss Factor, τ_a*	$\bar{\Delta T} = \Delta T \times (\tau_a)$	*Spatial Frequency*	*New Range, km*
9	0.155	0.387	14.0	10.50
10	0.126	0.315	13.6	10.20
11	0.102	0.255	13.2	9.90
12	0.083	0.207	12.8	9.60
13	0.068	0.170	12.4	9.30
14	0.054	0.135	12.1	9.07
15	0.045	0.112	11.7	8.78

10 km. Therefore, the recognition range of the thermal sensor will reduce from its value of about 13 km given in eq. (6.67) to 10 km, which includes atmospheric transmission losses.

6.10 SUMMARY

In this chapter, the system-oriented design and performance equations of electro-optical systems were derived and applied to numerical examples. New methods of including atmospheric attenuation characteristics in performance equations were developed and demonstrated. In writing this chapter, the best features of the corresponding section of the standard electro-optical reference books of Lloyd and Seyrafi, as well as the computer program of Ratches et al., were used. The analysis was carried beyond the contents of these references, in streamlining solutions for digital computers.

REFERENCES

[1] Hovanessian, S.A., *Radar System Design and Analysis*, Dedham, MA: Artech House, 1984, Chapter 6.

[2] Wolfe, W.L., and Zissis G.J., *The Infrared Handbook*, Ann Arbor: Environmental Research Institute of Michigan, 1978, p. 1–87.

[3] Kruse, P.W., L.D. McGlauchlin, and R.B. McQuistan, *Elements of Infrared Technology*, New York: John Wiley and Sons, 1962.

[4] Seyrafi, K., *Electro-Optical System Analysis*, Los Angeles: Electro-Optical Research Corporation, 1981, Chapter 7.

[5] Lloyd, J.M., *Thermal Imaging System*, New York: Plenum Press, 1975, Chapter 5.

[6] Obert, L.P., et al., *Visionics EO Sensor Performance*, Arlington, VA: DCS Corporation, June 1981.

[7] Ratches, J.A., et al., "Night Vision Laboratory Static Performance Model for Thermal Viewing Systems," R and D Technical Report, ECOM-7043, April 1975.

[8] Johnson, J., "Image Intensifier Symposium," Paper AD-220160, Fort Belvoir, VA, 1958.

[9] Hudson, R.D., *Infrared Systems Engineering*, New York: John Wiley, 1969.

[10] Hovanessian, S.A., *Introduction to Sensor Systems*, Norwood, MA: Artech House, 1988, Chapter 5.

Chapter 7
Electro-Optical System Applications

7.1 INTRODUCTION

Electro-optical system applications range from infrared search and tracking systems to visible and infrared imaging systems used in ground, air, and space applications. Visible imagers are used under daylight conditions whereas IRST systems, using target exhaust or temperature gradient for detection, can be used during the day or night. Infrared thermal imagers, usually referred to as *FLIR* (forward-looking IR), use 8–14 μm wavelength region of the spectrum; and they can produce images based on the temperature gradient of the imaged area during the day or night.

In this chapter several examples of IRST systems and visible and thermal imaging systems will be given both for scanning and staring systems. The examples will include numerical calculations illustrating the design of these systems.

7.2 INFRARED SEARCH AND TRACKING SYSTEMS

Infrared search systems operate much the same way as electro-optical thermal imagers. These systems consist of an optical telescope, scanning mechanisms, detectors, and signal processing equipment. Figure 7.1 shows an EO imaging system with a single detector. The object space resolution cell is imaged on the detector through horizontal and vertical scanners. Rotation of the horizontal scanner effectively moves the detected resolution cell across the horizontal object space, and the vertical scanner moves this point in the vertical plane of the object space. Thus the movement of scanners puts the entire scene through a single detector, producing the appropriate image on the display. In a multiple detector system, more than a single cell in the object space can be mapped into the detectors. This increases the S/N ratio of each resolution cell, enhancing the quality of the resulting image. In a *focal plane array* (FPA), a mosaic of detectors is used to map an entire scene. This is similar to photographic cameras, where the lens' field of view is produced on the film. Although

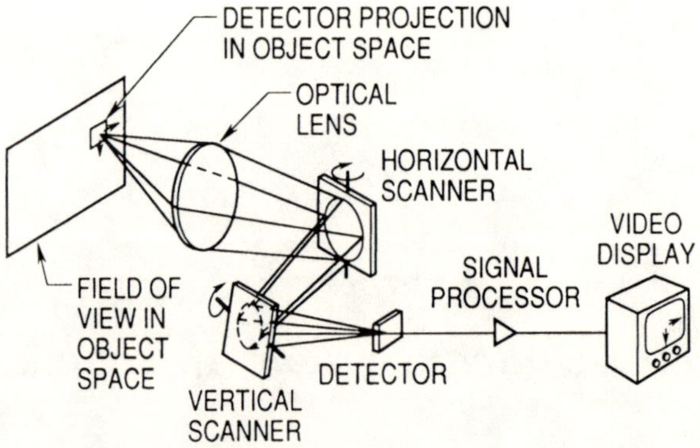

Figure 7.1 Electro-optical scanning sensors.

work is continuing toward entire scene registration using FPA concepts, it may be some time before these systems become operational.

The main difference between electro-optical thermal imagers and infrared search and tracking systems lies in the fact that electro-optical imagers are used to map ground areas or locate targets in the presence of ground reflection. Infrared search and tracking systems are usually used to locate targets in the clear or a sky background. These systems are utilized in ground-to-air and air-to-air applications. For example, infrared systems are used in a number of missile seekers to lock-on and track targets for missile guidance and target interception.

The electro-optical imaging and the infrared systems have the advantage of being passive devices where no transmission of energy occurs. This fact denies the enemy detection and advance warning. The radiation of active microwave radars, on the other hand, can be detected by the enemy, providing a valuable tactical advantage. The passive nature of EO imaging and IRST systems together with the current emphasis on stealth technology (whereby radar detection ranges are reduced due to reduced radar cross section of the targets) has greatly expanded the role of these systems [1]. Because infrared systems are passive devices containing only a receiver system, they usually weigh considerably less than active microwave systems with both a transmitter and a receiver.

The infrared wavelengths utilized in missile seekers are influenced by target radiation characteristics, atmospheric attenuation, and detector responsivity. Target radiation originates from hot metal parts of graybody radiation or combustion products of the engine. The combination of these effects reduces infrared system design to five specific IR bands: 1.5 to 1.8 μm, 2.0 to 2.5 μm, 3.5 to 4.2 μm, 4.5 to 4.8 μm, and 8.5 to 12.5 μm. Lead sulfide and lead selenide detectors are responsive in the 1.5–3.0 and 3.0–5.0 μm bands, respectively. These detectors were used in early

IR homing missiles. Development of cooling methods, such as cryogenic coolers that reduce detector temperatures to 70 or 80 K, allowed the utilization of mercury-cadmium-telluride detectors operating in the 8–12 μm region of IR wavelengths. These detectors are used in the night vision or FLIR systems.

7.3 INFRARED TARGET TRACKING

Infrared target trackers use a variety of methods for target tracking with an interesting application being utilization of reticles [2]. Consider the optical system of Figure 7.2 showing target rays from a boresighted target, as well as, rays from an off-boresight target. These rays are reflected from the primary mirror onto a secondary mirror, converging at the detector. Placement of a "wagon wheel" reticle (Figure 7.3) in front of the secondary mirror will modulate the incoming signal.

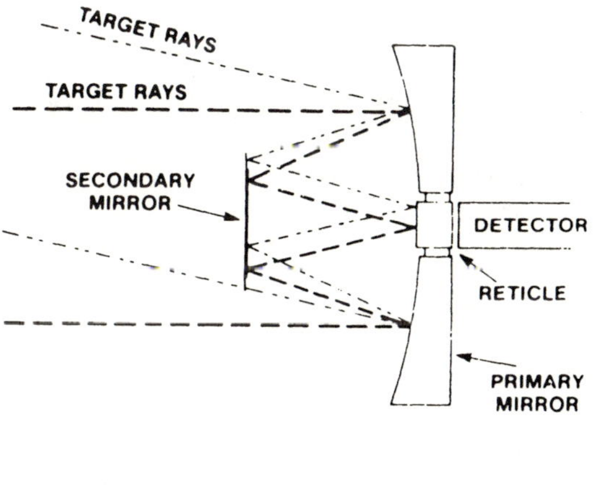

Figure 7.2 Optical ray traces from boresight and off-boresight targets.

For the boresight target of Figure 7.3, the spinning wheel will produce equal durations of "ones" and "zeros" as opaque and transparent portions of the wheel intercept the target energy in equal durations. The off-boresight target of Figure 7.3 will produce unequal durations of target energy impinging on the detector. This is because unequal lengths of opaque and transparent portions of the wheel intercept the incoming energy.

The frequency modulation of target return of Figure 7.3 is used through proper circuitry to boresight the IR set with the target and subsequent target tracking.

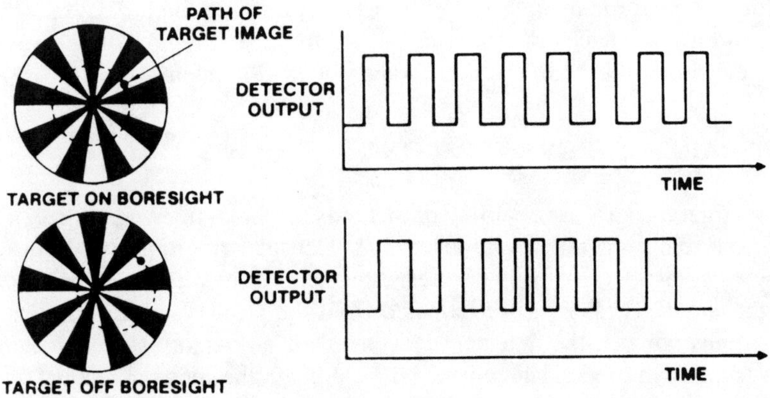

Figure 7.3 Wagon wheel reticle and target returns.

Another method of IRST single target tracking is by means of quadrant tracker [3]. Figure 7.4 shows four detectors A, B, C, and D with the defocused spot representing the target return energy. The output of each detector is proportional to the image power incident upon its sensitive area. The error signal for elevation and

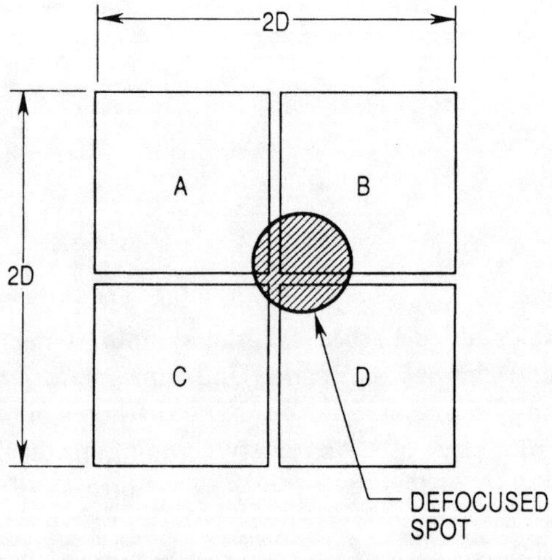

Figure 7.4 Quadrant detectors of an electro-optical tracker.

azimuth axis will be

$$\varepsilon_{el} = \frac{(I_A + I_B) - (I_C + I_D)}{I_A + I_B + I_C + I_D} \tag{7.1}$$

$$\varepsilon_{az} = \frac{(I_B + I_D) - (I_A + I_C)}{I_A + I_B + I_C + I_D} \tag{7.2}$$

where I_x is the peak signal current of detector x, and the denominators of equations (7.1) and (7.2) represent the sums of the signal currents.

The signal processing portion of the tracking system must operate over a wide dynamic range. In general, the dynamic range of linear amplification may not be sufficient, and some form of logarithmic amplification is usually needed. A block diagram of a typical electro-optical signal processing system is given in Figure 7.5.

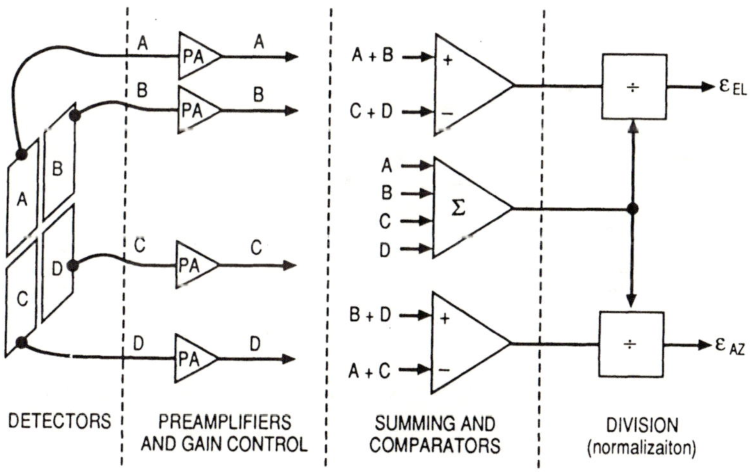

Figure 7.5 Signal processing diagram of a quadrant electro-optical tracker.

Passive IRST systems provide accurate angular position information but they do not provide target range. For this reason, these systems can be used in conjunction with laser range finders. In a typical application the IRST searches a given volume of the search area and locates the target in angular position. The laser range finder is then pointed in the direction of the target to obtain range and, also, to confirm the existence of the target as opposed to false alarm.

7.4 EXAMPLES OF ELECTRO-OPTICAL SCANNING AND STARING SENSORS

In this section four examples of electro-optical systems will be given. These examples are of a shipboard IRST system, an airborne IRST system, a narrow field missile seeker IRST, and threat warning receiver IRST sensor.

7.4.1 Shipboard IR Search and Track System

Consider the shipboard IRST sensor illustrated in Figure 7.6. The purpose of this sensor is to detect the incoming targets at long ranges and track them continuously. A linear scanning array with the possibility of TDI is considered. The arrays will be HgCdTe detectors operating in the MWIR region of the spectrum at temperatures of 60–80 K. The sensor has a 10-in diameter and a focal number of unity. Spatial resolution is to be 0.35 mrad. The sensor should cover a field of view of 360° by 20° in 1 s frame time. It is required to detect a 10 W/sr target at a range of 50 km.

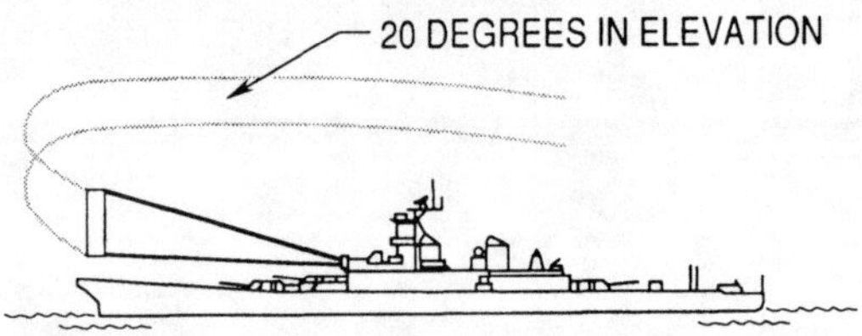

Figure 7.6 Wide angle, high-resolution scanning sensor for shipboard application.

Using the S/N ratio of eq. (6.13), we utilize the following notation-parameter combinations:

J_t	10 W/sr	
R	50 km	
FOV	360 × 20	
T_f	1 s	
D	10 in	
$f/\#$	1	(7.3)
mode	scanning	
D^*	10^{11} cm-$H_z{}^{1/2}$/W	
wavelength	3–5 μm	
$\alpha_d^{1/2}$	0.35 mrad	
array size	1,000	

where the value of D^* is obtained from Figure 6.2. Assume an atmospheric loss factor of 0.30. The following loss parameters are also assumed for eq. (6.14):

$$K_o = 0.7; \quad K_e = 0.9; \quad K_d = 1; \quad K_m = 1; \quad K_n = 1; \quad \text{and } K_{fd} = 1.2 \tag{7.4}$$

Using these values, the loss factor becomes

$$\begin{aligned} L &= K_o K_e K_d K_m (K_n / K_{fd}) \\ &= (0.7)(0.9)(1)(1)(1/1.2)^{1/2} \\ &= 0.575 \end{aligned} \tag{7.5}$$

Using the S/N equation from Chapter 6 as follows

$$S/N = \frac{\pi D J_t D^* L \tau_0}{4(f/\#)R^2} \left(\frac{nT_f}{\Omega_s} \right)^{1/2} \tag{7.6}$$

with parameters of eqs. (7.4) and (7.5), we get

$$\begin{aligned} S/N &= \frac{\pi(10 \times 2.54)(10)(10^{11})(0.575)\,(0.3)}{4(1)(50E5)^2} \left(\frac{(1000)(1)}{2.1929} \right)^{1/2} \\ &= 2.939 \\ &= 4.68 \text{ dB} \end{aligned} \tag{7.7}$$

where the Ω_s value of 2.1929 sr representing a FOV of 360° by 20° is used in eq. (7.7). The elevation field of view of 20° is obtained from 1,000 detectors of 0.35 mrad each:

$$1000 \times 0.35 \times 0.0573 = 20.0° \tag{7.8}$$

Therefore, the array of 1,000 detectors in elevation can be scanned in azimuth for 360° per second as shown in Figure 7.7.

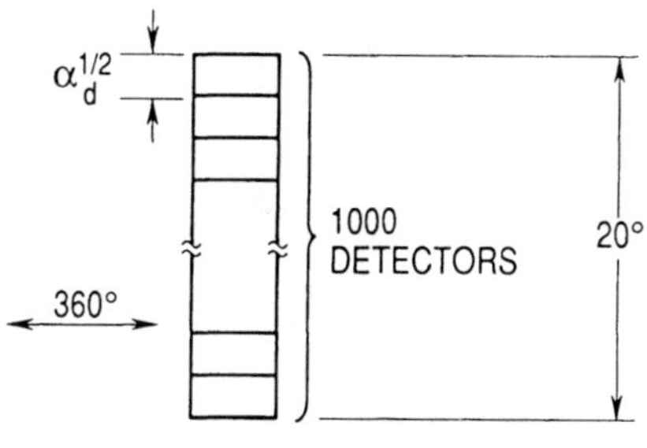

Figure 7.7 Array of 1,000 vertical detectors scanning 360° in azimuth.

To improve the S/N ratio, time delay integration can be used. For example a string of 16 detectors in TDI format will increase the S/N ratio by a factor of 4 or 6 + B to a value of 10.68 dB.

The number of bits required to transfer the data to another signal processing or display station can be computed by calculating the number of resolution cells scanned by the IRST system. The number of resolution cells (pixels) is

$$n = \frac{\Omega_s}{\alpha_d}$$

$$= \frac{2.193}{(0.35 \times 10^{-3})^2} \quad (7.9)$$

$$= 17.9E6 \text{ pixels}$$

Because the scanning time is 1 s this number of pixels should be covered in 1 s. Assuming 12 bits per pixel, the number of bits per second becomes

$$\text{BPS} = 215\,E6 \quad (7.10)$$

7.4.2 Airborne IR Search and Track System

An airborne IRST sensor is shown in Figure 7.8. The purpose of this sensor is to detect and track large targets. The sensor uses 580 × 4, scanning TDI arrays of HgCdTe detectors operating in the LWIR region (8–10 μm) with an operating temperature of 60–80 K. The sensor has 8 in optics with an f number of 2, a resolution of 0.3 mrad, and covers a FOV of 360° × 10° in 4 s. Assume a target of 10 W/sr at a range of 20 km. For this system use the loss values of eq. (7.5) and atmospheric attenuation factor τ_0 of 0.3. The value of D^* for LWIR (8–10 μm) can be obtained

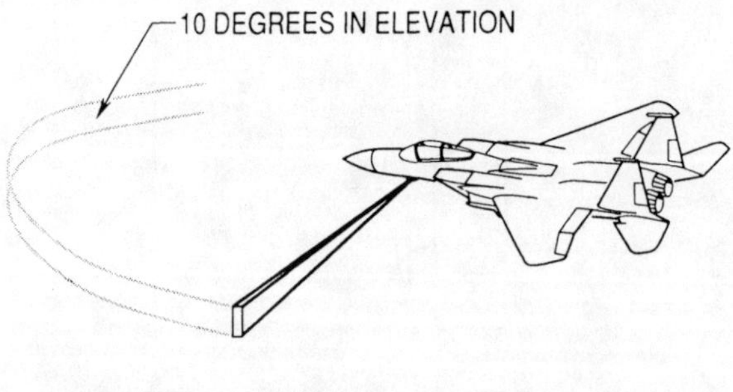

Figure 7.8 An airborne IRST for wide area search.

from Figure 6.2 as 2×10^{10} cm-Hz$^{1/2}$/W. Using these values in the S/N ratio equation (7.6) we get

$$\frac{S}{N} = \frac{\pi(8 \times 2.54)(10)(2 \times 10^{10})(0.575)(0.30)}{4(2)(20E5)^2}\left(\frac{(580 \times 4 \times 4)}{1.096}\right)^{1/2}$$

$$= 6.33 \tag{7.11}$$

$$= 8 \text{ dB}$$

Note that the elevation coverage of this sensor with 580 detectors will be

$$(580)(0.30)(0.057) = 10° \tag{7.12}$$

Using this, the azimuth coverage can be obtained by rotation of the detectors 360°.

7.4.3 Small Aperture Staring Array IRST

Consider a small field of view, high-resolution, staring array detection and tracking sensor. The sensor utilizes 90 × 90 hybrid of HgCdTe detectors operating in MWIR (3–5 μm) region with an operating temperature of 60–80 Kelvin. The sensor has a 2-in diameter optics and covers a field of view of 2.5° by 2.5°. Assume a target of 10 W/sr at 20 km range and an optical f number of 2.

The detector FOV is 0.48 mrad and the atmospheric attenuation factor is 0.1. Assume that focal plane array is scanned at a rate of 100 frames per second.

Using the number of detectors and the instantaneous detector FOV we have the angular coverage in azimuth and elevation as

$$(90)(0.48)(0.057) = 2.5° \tag{7.13}$$

This value was given in the problem.

Although the S/N ratio equation (7.6) was written for a scanning sensor, it can be applied equally well for the preceding starring array. Using eq. (7.6), we have

$$\frac{S}{N} = \frac{\pi(2 \times 2.54)(10)(10^{11})(0.5)(0.1)\left(\frac{90 \times 90 \times 1}{(100)(1.9E-3)}\right)^{1/2}}{4(2)(20E5)^2}$$

$$= 5.14 \tag{7.14}$$

$$= 7.116 \text{ dB}$$

where a D^* value, from Figure 6.2, of 10^{11}, and a loss factor of 0.5 is used.

The number of pixels in the field of view, number of pixels per second, and the number of bits per second will be

$$n = \frac{(2.5)(2.5)}{(57.3)^2(0.48E-3)^2} = 8.26E3 \text{ pixels}$$

$$\frac{8.26E3}{1/100} = 8.26E5 \text{ pixels/s} \tag{7.15}$$

$$8.26E5 \times 12 = 9.91E6 \text{ bits/s}$$

where 12 bits is assumed per pixel S/N ratio.

7.4.4 Step-Stare Detection System

A small aperture, wide field of view IR detection and tracking system is shown in Figure 7.9. The sensor uses 128 × 128 staring arrays of MWIR HgCdTe detectors operating between 60 and 80 K. The sensor has 0.5-in diameter optics with an *f*/1.5 lens system. The resolution of the detector is 4 mrad and the sensor covers a 60° × 60° FOV. A target with 10 W/sr is at a 20 km range. Compute the S/N ratio.

The IRST system parameters are given as;

J_t	10 W/sr	
R	20 km	
FOV	60 × 60	
D	0.5 in	
T_f	1 s	
$\alpha_d^{1/2}$	4 mrad	(7.16)
$f/\#$	1.5	
mode	step-stare	
array size	128 × 128	
D^*	10^{11} cm-Hz$^{1/2}/W$	
wavelength	3–5 μm	
τ_a	0.3	

The detector array of 128 × 128 will result in a 30° by 30° FOV as follows:

$$(128)(4E-3)(57.3) = 30° \tag{7.17}$$

Therefore, we would need four sets of 128 × 128 detectors for a 60° × 60° coverage as shown in Figure 7.10. Considering the frame time of 1 s and assuming a scanning

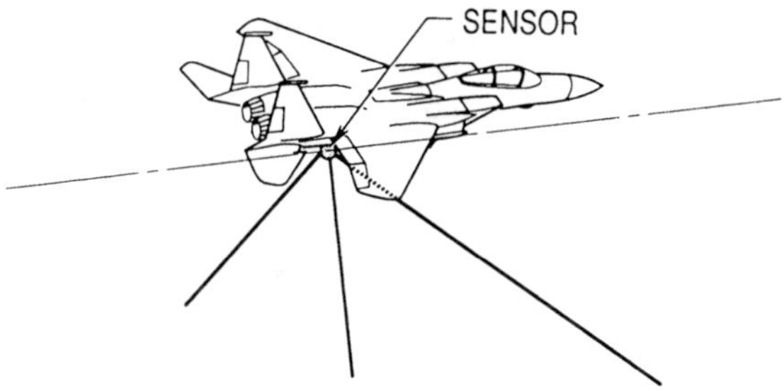

Figure 7.9 Threat warning sensor.

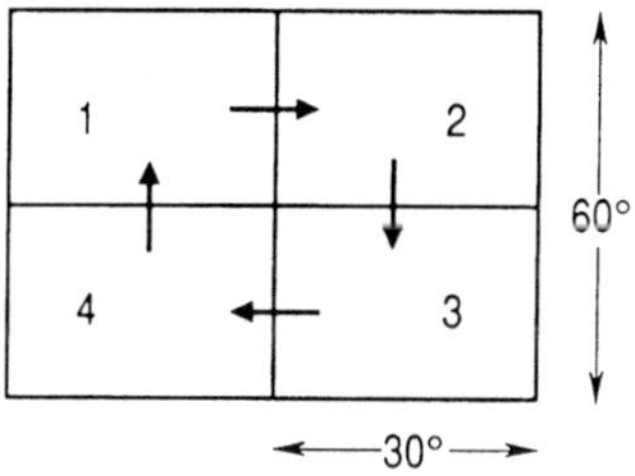

Figure 7.10 Step and stare detection arrangement of IRST detectors for a 60° × 60° coverage.

efficiency of 80%, the frame time for each one of the four step and stare steps will be

$$T_f = 0.2 \text{ s} \tag{7.18}$$

With these values we can calculate S/N ratio of eq. (7.6) as follows:

$$\frac{S}{N} = \frac{\pi(0.5 \times 2.54)(10)(10^{11})(0.5)(0.3)}{4(1.5)(20E5)^2}\left(\frac{(128 \times 128 \times 0.2)}{(0.274)}\right)^{1/2}$$

$$= 2.726 \tag{7.19}$$

$$= 4.36 \text{ dB}$$

where a loss factor of 0.5 and an equivalent Ω_s of 0.274 sr is used. The number of pixels in the FOV will be 6.8E4 and the number of bits per second will be 8.22E5.

7.4.5 False Alarm Rates and Threshold Setting

The signal-to-noise ratios just calculated should be used in conjunction with false alarm rates and threshold settings as discussed in Chapters 5 and 6. The false alarm rate is given by eq. (6.18) as

$$\text{FAR} = \frac{P_n \Omega_s}{\alpha_d T_f} K_n \tag{7.20}$$

where FAR is the false alarm rate per second, P_n is the probability of false alarm, Ω_s is scan volume in steradians, α_d is the detector instantaneous FOV, T_f is the scan frame time, and K_n is the scanning efficiency. The probability of false alarm P_n is given in terms of threshold setting E_t as

$$P_n = \exp\left(-\frac{E_t^2}{2}\right) \tag{7.21}$$

Combining eqs. (7.21) and (7.20) we can solve for threshold setting in terms of false alarms rate:

$$\frac{E_t^2}{2} = -\ln\left(\frac{\alpha_d}{\Omega_S} \frac{T_f \text{FAR}}{K_n}\right) \tag{7.22}$$

where ln represent natural logarithm. For a given false alarm rate, and other parameters, eq. (7.22) will give the threshold-to-noise ratio. Note that E_t represents the peak threshold to rms noise voltage value and $E_t^2/2$ represents rms threshold power-to-rms noise power value [4].

As a numerical example of the false alarm rate calculations, consider the parameters of the example given by eq. (7.16): that is, the step-stare detection system. Assuming we can tolerate one false alarm in each quadrant of Figure 7.10 every hour, we have from eq. (7.22)

$$\begin{aligned} \frac{E_t^2}{2} &= -\ln \frac{(4E-3)^2(0.2)(1/3600)}{(0.274)(1)} \\ &= 19.5 \\ &= 12.9 \text{ dB} \end{aligned} \tag{7.23}$$

where a scanning efficiency K_n of 1 is assumed. This value of 12.9 dB, from Figure 6.5 or eq. (7.21), corresponds to a probability of false alarm of $P_n = 10^{-9}$. Using the corresponding detection curve for this value of P_n we see that the S/N ratio obtained in eq. (7.19) is completely inadequate for the selected false alarm rate. To obtain the false alarm rate given earlier, the design should be changed completely. With the present design only at a range of 5 km, the S/N will be adequate for detection. That is, at 5 km the S/N of eq. (7.19) will be 16.36 dB. Using this value and the $P_n = 10^{-9}$ curve of Figure 6.5, we obtain a probability of detection of over 95%.

7.5 SPACEBORNE IMAGING SYSTEMS

Spaceborne electro-optical imaging systems have been used for Earth observation for about two decades. These systems use visible and IR spectrum to obtain information as to land use, ocean studies, and weather and cloud coverage information. In this section several spaceborne imaging systems currently operational or planned for future deployment will be discussed.

7.5.1 Landsat System of Satellites

The Landsat system of satellites using electro-optical imaging devices have been operational, although not continuously, since July 23, 1972, when *Landsat 1* was launched. The satellites are utilized to obtain images of the Earth for a variety of studies including agriculture, forestry, land use and mapping, geology, water resources, oceanography, and the environment.

The latest in the series of landsat satellites is *Landsat 6* (L6), which will be launched into a 705-km, sun-synchronous orbit in mid-1992 [5]. Figure 7.11 gives a view of the satellite as it will be deployed and also a cut-away of enhanced thematic mapper (ETM). Figure 7.12 is a description of the ETM orbital geometry and of its ground coverage. The ETM weight is about 827 lbs and the power consumption is about 400 W. Landsat will employ two high data rate *Landsat recording systems* (LRS) units for onboard data storage and playback. Each unit is capable of recording and reproducing at a rate of 85 Mbits per second and can store up to 75 Gbits.

The electro-optical sensor of the system (ETM) will operate in eight spectral bands as given in Table 7.1. The ETM in addition to having visible (blue, green, and red) and infrared bands, will have a panchromatic band in the range of 0.50 to 0.90 μm. Table 7.1 gives the *instantaneous ground field of view* (IGFOV) as well as ground resolution, about 30 m, for each band.

As discussed in Chapter 4, the spectral bands are selected to give specific information about the imaged area. Table 7.2 gives spectral band characteristics of the ETM. These characteristics include vegetation health, water content, cloud discrimination and so on. Figures 7.13 and 7.14 give thematic mapper images obtained from previous space flights. Figure 7.13 is the map of several areas of Washington, DC,

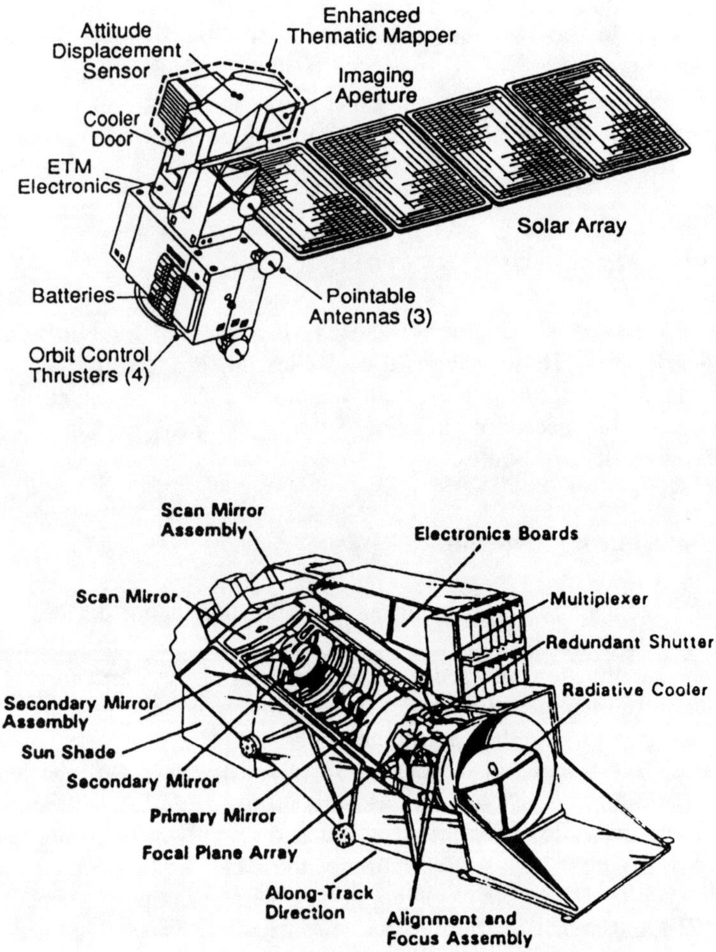

Figure 7.11 (a) Fully deployed on-orbit Landsat 6 configuration and (b) cutaway drawing of ETM scanner.

where vegetation (green, band 2), chlorophyll absorption (red, band 3), and water delineation (band 4) information is clearly shown. Figure 7.14 shows visible bands 2, 3, and 4, where snow and cloud discrimination is impossible, and also bands 2, 5, and 7, where clouds are clearly distinguished.

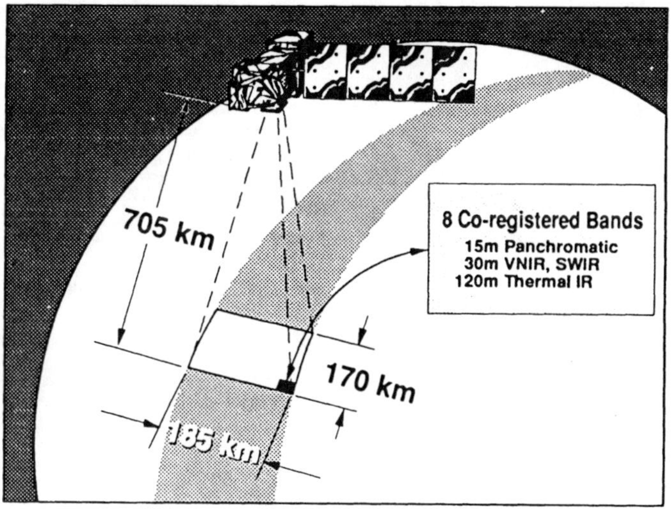

Figure 7.12 ETM scanning geometry.

Planned future moderate- and high-resolution spectrometers, as will be discussed in the next section, have even finer spectral bands than ETM systems. These systems are designed to extract greater land use information by using spectral analysis

Table 7.1
Summary of Eight ETM Spectral Bands

Band #	*Bands (μm)*	*Detectors*	*IGFOV (μrad)*	*Ground Resolution (m)*
PAN	0.50–0.90	SiPD (32)	18.5 × 21.3	13 × 15
1	0.45–0.52	SiPD (16)	42.5	30
2	0.52–0.60	SiPD (16)	42.5	30
3	0.63–0.69	SiPD (16)	42.5	30
4	0.76–0.90	SiPD (16)	42.5	30
5	1.55–1.75	InSb (16)	42.5	30
7	2.08–2.35	InSb (16)	42.5	30
6	10.4–12.5	HgCdTe (4)	170	120

Table 7.2
Landsat Thematic Mapper Spectral Band Characteristics

Band	*Range (μm)*	*Characteristic*
1	0.45–0.52	Coastal water mapping
2	0.52–0.60	Healthy vegetation
3	0.63–0.69	Chlorophyll absorption
4	0.76–0.90	Water delineation, vegetation vigor
5	1.55–1.75	Snow or cloud differentiation
6	10.4–12.5	Plant-heat stress measurement
7	2.08–2.35	Hydrothermal mapping

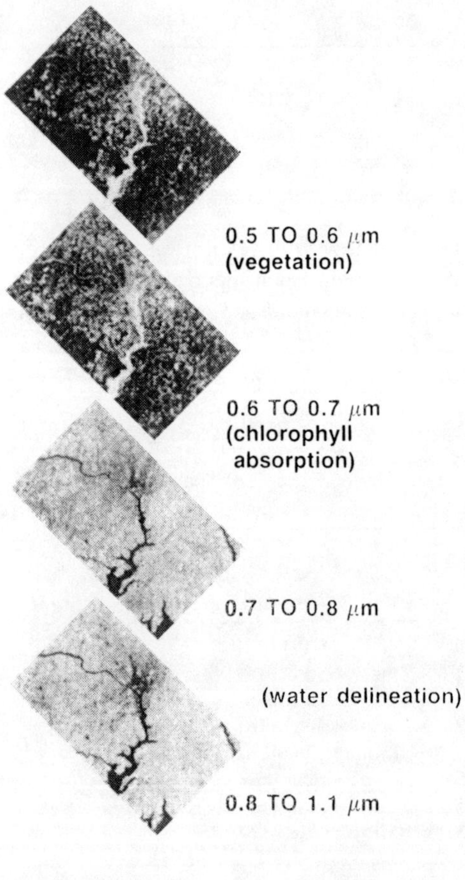

Figure 7.13 Maps of the Washington, DC area in several spectral bands.

(a) (b)

Figure 7.14 Snow-cloud discrimination: (a) bands 2, 3, and 4, no cloud-snow separation; (b) bands 2, 5, and 7, clouds stand out and snow is eliminated.

7.5.2 Moderate-Resolution Imaging Spectrometer

The *moderate-resolution imaging spectrometer* (MODIS) is a follow-up to Landsat thematic mapper for large spatial scale ($\geq$ 1 km) observation requirements within the atmosphere [6]. This system is designed to be launched as part of the *Earth Observing System* (EOS) in 1997. It has 64 spectral bands covering the spectral region of 0.410 to 1.040 μm.

As an example of S/N ratio calculation of this sensor consider the geometry of Figure 7.15 and the following sensor parameters:

Satellite altitude	705 km	
Satellite ground velocity	6.76 km/s	
Ground resolution	1,000 m	
Swath width	1513 km	
Wavelength	0.41–1.040 μm	
Spectral bandwidth	10 nm	
Solar zenith angle	22°	(7.24)
Sensor look angle	20°	
Optical transmission	0.1	
Detector size	104 μm	
Telescope diameter	5 cm	
Optical $f/\#$	1.5	
Number of detectors per spectral band	64	
Scanning efficiency	0.85	

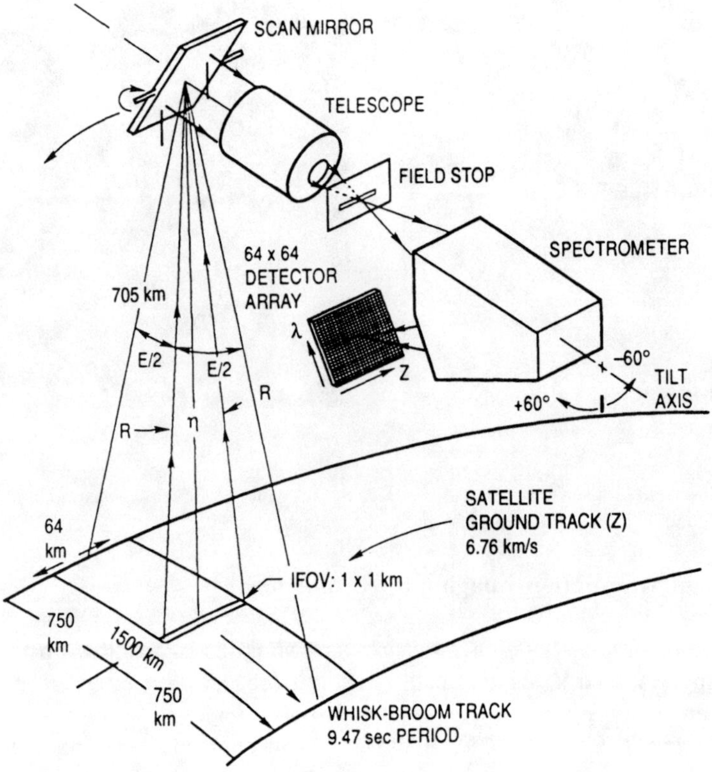

Figure 7.15 MODIS-T scan geometry and conceptual system layout.

Assume that a silicon detector is used at 300 K and calculate the S/N at 0.5 μm against a vegetation background.

The appropriate S/N ratio equation is given as in eq. (7.6). Most of the parameters of this equation are given in eq. (7.24) others, however, need to be calculated.

The radiation intensity at 0.50 μm and spectral bandwidth of 0.010 μm or 10 nm can be calculated with information provided in Figure 7.16 for a vegetated area [7]:

$$\begin{aligned} J_t &= (5.2E - 3)\,(1000)^2(100)^2(0.01) \\ &= 0.520E6 \text{ W/sr} \end{aligned} \tag{7.25}$$

where the first number represents radiance from Figure 7.16, the second number is the resolution in meters, the third number is meter-centimeter conversion, and the

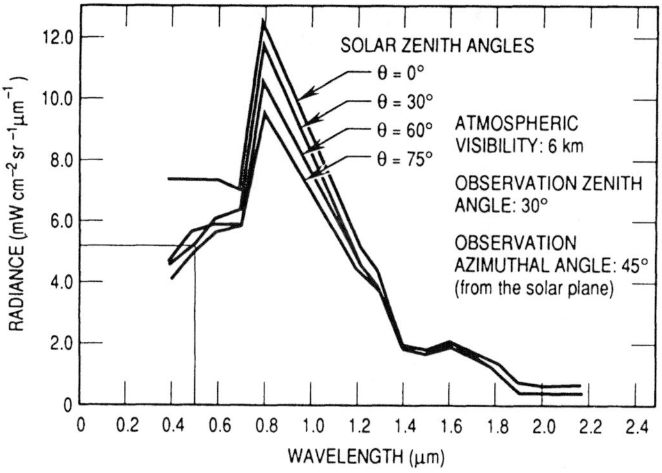

Figure 7.16 Upwelling spectral radiance observed at the top of a plane parallel with Earth's surface composed of vegetation.

fourth number is the spectral bandwidth. The sensitivity of the detector can be obtained from Figure 7.17 [3], at a wavelength of 0.50 μm, as

$$D^* = 1.3E12 \text{ cm-H}_z{}^{1/2}/\text{W} \tag{7.26}$$

The loss factor can be computed from the given values as

$$L = (0.85)(0.1) = 0.085 \tag{7.27}$$

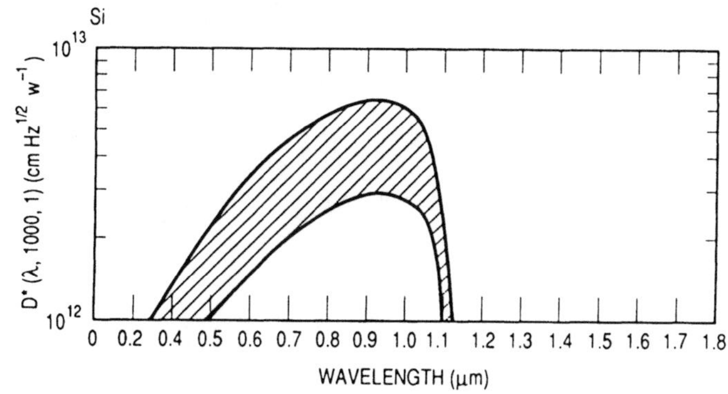

Figure 7.17 Silicon detector performance at 300 K.

The number of detectors as given in eq. (7.24) is 64, and the frame time as illustrated in Figure 7.15 is 9.47 s. Note that this is also the time that the satellite moves a distance of 64 km with a satellite velocity of 6.76 km/s.

The angular coverage of the sensor can be computed from range computation of Figure 7.15,

$$\begin{aligned} R &= \sqrt{(750^2 + 705^2)} \\ &= 1029 \text{ km} \end{aligned} \tag{7.28}$$

and the elevation coverage angle will be

$$\begin{aligned} \frac{\in}{2} &= \arctan \frac{750}{705} \\ \in &= 93.5^\circ \end{aligned} \tag{7.29}$$

The azimuth coverage will be

$$\begin{aligned} \frac{\eta}{2} &= \arcsin \frac{32}{705} \\ \eta &= 5.2^\circ \end{aligned} \tag{7.30}$$

Using these values,

$$\begin{aligned} \Omega &= \frac{\in \eta}{(57.3)^2} \\ &= 0.148 \text{ sr} \end{aligned} \tag{7.31}$$

Using the calculated values in the S/N ratio equation (7.6) we get

$$\begin{aligned} \frac{S}{N} &= \frac{\pi(5)(0.52E6)(1.3E12)(0.085)}{4(1.5)(1029E5)^2} \left(\frac{64 \times 9.47}{0.148} \right)^{1/2} \\ &= 909 \\ &= 29.5 \text{ dB} \end{aligned} \tag{7.32}$$

Certainly this is a sufficient S/N ratio for detection and mapping.

Using the given values, a number of other informative calculations can also be made. Using eq. (6.6), we have

$$A_d = \alpha_d \, [(f/\#) \, D]^2 \tag{7.33}$$

where A_d and α_d are detector area and angular coverage, respectively. Using the given values of eq. (7.24), we have

$$(104E - 6)^2 = \alpha_d[(1.5)(0.05)]^2$$

$$\alpha_d = 1.92E - 6 \tag{7.34}$$

$$\alpha_d^{1/2} = 1.38 \text{ mrad}$$

The diffraction limit of the lens γ_d at 0.50 μm will be

$$\gamma_d = \frac{2.44\lambda}{D}$$

$$= 24.4 \ \mu\text{rad} \tag{7.35}$$

Note that the diffraction limit of the lens γ_d is considerably less than the FOV of the detector. Therefore, the target radiance passing through the lens will impinge on the detector with the detector determining the final system resolution.

The detector FOV of 1.3 mrad will represent a distance of 0.916 km at 705 km and a distance of 1.33 km at a range of 1,029 km at the extreme of the coverage. These values should be corrected to the nominal 1 k resolution range.

7.5.3 High-Resolution Imaging Spectrometer

The *high-resolution imaging spectrometer* (HIRIS) is another electro-optical sensor in the Earth Observing System of instruments for study of Earth's environment [8]. This sensor is designed to acquire images in 192 spectral bands covering the region of 0.4 to 2.5 μm, at a spectral sampling internal of 10 nm. The instantaneous ground field of view will be 30 m over a 30 km swath. A pointing capability will allow image acquisition up to +60° and −30° down track and ±24° or more cross track. The cross-track pointing will allow multiple viewing opportunities during one orbital revisit cycles, nominally 16 days.

As an example of S/N ratio calculation for HIRIS system consider the following orbital parameters from [8]:

Nominal altitude	825 km	
Orbital period	101 min	
Inclination (with respect to the equator)	98.7°	
Revolutions per day	14.2	(7.36)
Successive orbit ground track spacing	2824.7 km	
Ground track speed at the equator	7.383 km/s	
Adjacent track separation at the equator	176.5 km	

The approximate values of these orbital parameters can be obtained from the following equations [9]. The equations are given so that the reader may have an idea of obtaining these parameters, they are not meant to be complete or very accurate. The orbital velocity and period are

$$V_s = 7.9\left(\frac{R_e + h}{R_e}\right)^{-1/2}$$

$$T = \frac{2\pi(R_e + h)}{V_s} \tag{7.37}$$

$$V_e = \omega R_e$$

where

V_s is the velocity of the satellite, km/s,
R_e is the radius of Earth, km,
h is the height of the satellite, km,
T is the orbital period, s,
V_e is the velocity of Earth, km/s,
ω is the rate of rotation of Earth, rad/s.

Using an Earth radius of 6,372 km and an altitude of 825 km, the values of V_s and T can be obtained as

$$\begin{aligned} V_s &= 7.43 \text{ km/s} \\ T &= 101.66 \text{ min} \end{aligned} \tag{7.38}$$

Considering a 24-hr rotation period for Earth, we have

$$V_e = \frac{2\pi(6372)}{(24)(60)(60)} = 0.46338 \text{ km/s} \tag{7.39}$$

Using this, successive ground spacing of the satellite orbit will be

$$\text{Ground spacing} = (T)(V_e) = 2826.57 \text{ km} \tag{7.40}$$

Figure 7.18 gives the orbital parameters and the direction of satellite motion. Note that this is a sun-synchronous north-south orbit with ground position changing as a function of days and satellite orbits.

The optical layout of the HIRIS is given in Figure 7.19. The focal plane assembly includes the spectrometer entrance slit and VNIR and SWIR detectors.

The following sensor parameter are also given in [8]:

Lens aperture	37.6 cm	
Focal number	3.8	
Spectral band	0.4–2.5 μm	
Spectral bandwidth	0.01 μm	
Sensor FOV	2.1°	(7.41)
Detector IFOV	36 μrad	
No. of spatial detectors	1,000	
No. of spectral detectors	128	

Calculate the S/N ratio in the optical region of 0.5 μm for a bandwidth of 0.01 μm using a silicon detector. The detector sensitivity from Figure 7.17 will be

$$D^* = 10^{12} \text{ cm-Hz}^{1/2}/\text{W} \tag{7.42}$$

The ground resolution value ΔR, using the detector IFOV of 36 μrad and an altitude of 825 km, will be

$$\Delta R = (825E3)(36E - 6) = 30 \text{ m} \tag{7.43}$$

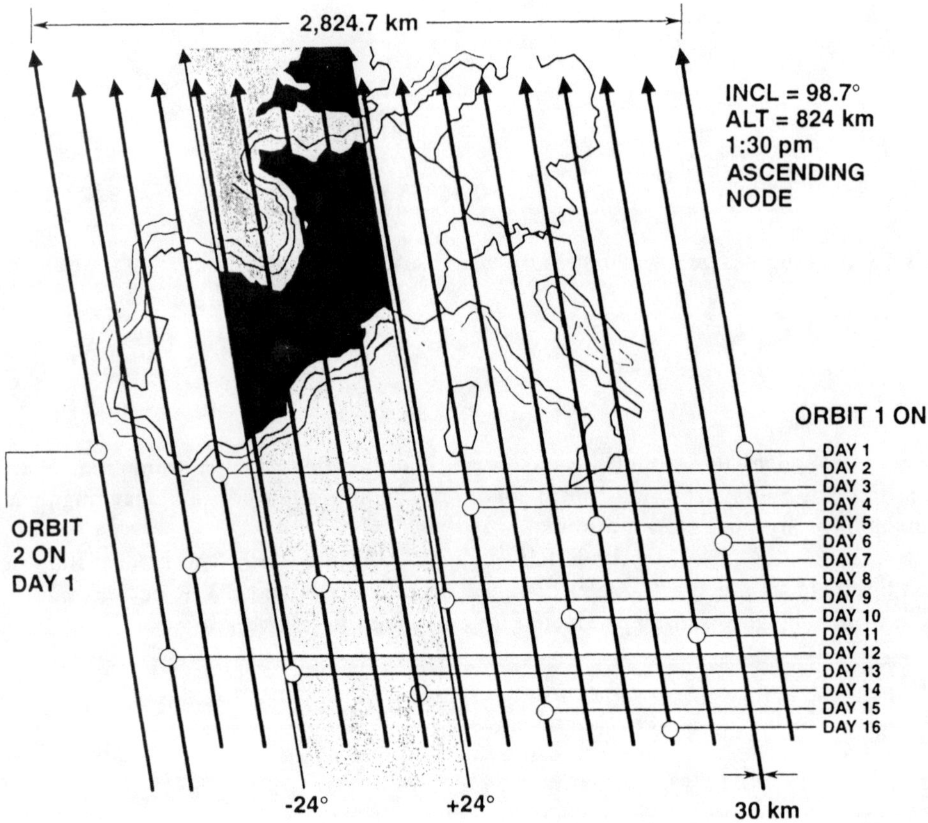

Figure 7.18 HIRIS orbital coverage pattern (shaded area represents possible coverage on a given orbit).

Consider the radiance of soil and rock as illustrated in Figure 7.20. A value of 8 mW per cm^2-sr-μm seems reasonable from this figure. Using this, the radiated power from each resolution cell will be

$$J_t = (30 \times 30)(100)^2 (8E - 3)(0.01)$$
$$= 720 \text{ W} \quad (7.44)$$

where the last term is the spectral bandwidth of the detector. The loss factor in this bandwidth is assumed to be 40% or

$$L = 0.40 \quad (7.45)$$

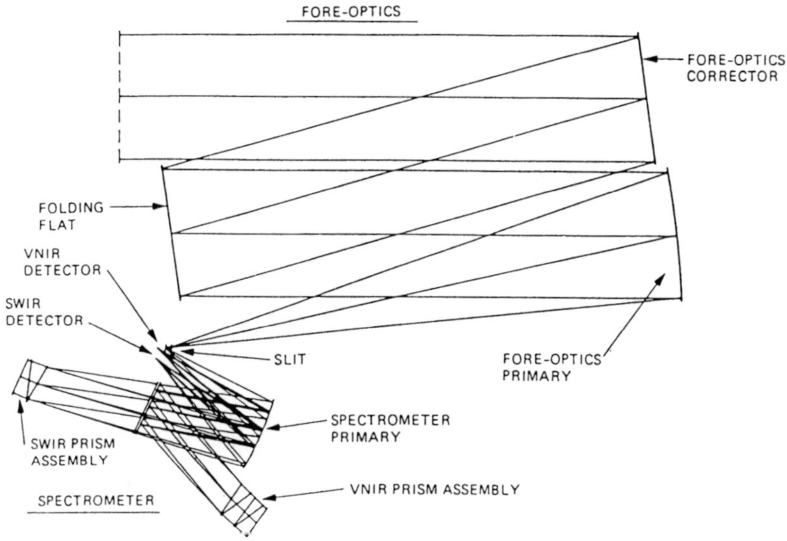

Figure 7.19 HIRIS optical layout.

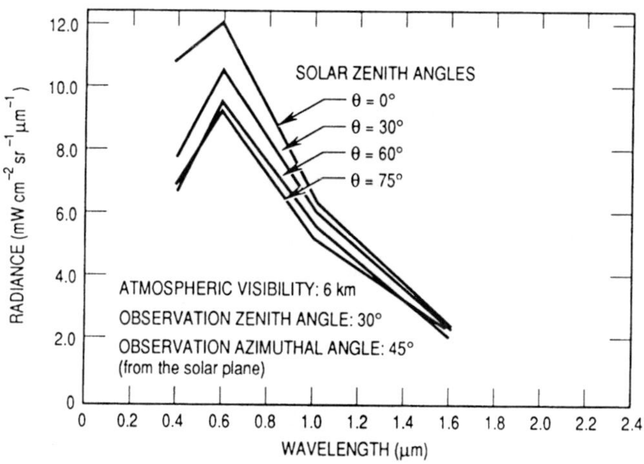

Figure 7.20 Upwelling spectral radiance observed at the top of a plane parallel to Earth's surface composed of soil and rocks. (From [7].)

The ground coverage, G_c, as seen from Figure 7.21 is

$$G_c = (2.1)(825)/57.3$$
$$= 30235 \text{ m} \tag{7.46}$$

Because we have 1,000 spatial detectors, the spatial resolution using eq. (7.46) will be 30 m, as stated previously. Assuming that we cover 6.667 km/s in the flight direction, the angular coverage per second will be

$$\text{Angular coverage} = \frac{6.667}{825} \times 57.3$$
$$= 0.463° \tag{7.47}$$

Using this coverage with 2.1° coverage in the perpendicular direction we have

$$\Omega_s = \frac{(2.1)(0.463)}{(57.3)^2}$$
$$= 2.961E - 4 \text{ sr} \tag{7.48}$$

Using the parameters given previously in the S/N ratio equation (7.6) we get

$$\frac{S}{N} = \frac{\pi(37.6)(720)(10^{12})(0.40)}{4(3.8)(825E5)^2}\left(\frac{(1000)(1)}{2.961E - 4}\right)^{1/2}$$

$$= 604 \tag{7.49}$$

$$= 27.8 \text{ dB}$$

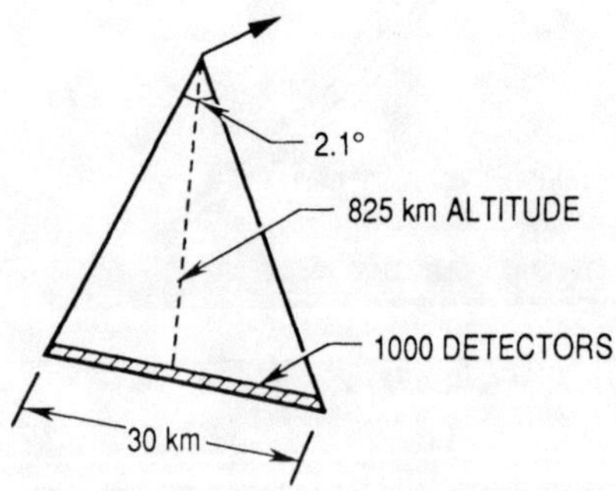

Figure 7.21 Flight geometry and ground coverage.

Note that this S/N ratio does not include optics losses. Even with the inclusion of possible 10 dB optical transmission loss the S/N ratio will be a respectable 17.8 dB.

REFERENCES

[1] "Emphasis on Stealth Will Spur Improvements on FLIR Sensors," *Aviation Week and Space Technology*, September 7, 1987.

[2] May, J.J., and M.E. Van Zee, "Electro-Optical and Infrared Sensors," *Microwave J.*, September 1983, p. 121–131.

[3] Wolfe W.L., and G.J. Zissis, *The Infrared Handbook*, Ann Arbor: Environmental Research Institute of Michigan, 1978, p. 22–87.

[4] Hovanessian, S.A. *Introduction to Sensor Systems*, Norwood, MA: Artech House, 1988, p. 205.

[5] Mowle, E.W. and C.J. Donnehy "The Landsat-6 Satellite: an Overview," IEEE AES Magazine, June 1991, p. 18.

[6] "*Earth Observing System Reports*," Volume IIb, MODIS, NASA, 1986.

[7] Rose, H, et al., *The Handbook of Albedo and Thermal Earthshine*, Ann Arbor: Environmental Research Institute of Michigan, Report No. 190201-1-T, 1973.

[8] "*Earth Observing System Reports*," Volume IIc, HIRIS, NASA, 1987.

[9] Chobotov, V. (Ed.), *Orbital Mechanics*, AIAA, 1991.

Chapter 8
Laser Radar Systems

This chapter contains a general discussion of the design and performance equations of laser radar systems. The signal power and noise power equations are derived for both direct and heterodyne laser radar systems. Probability of detection curves are given for glint and speckled targets, and methods of calculating the number of false alarms are discussed.

Ground backscattering and natural interference signals—such as emittance from sky, clouds, and ground—are discussed, and methods of incorporating these signals in detection calculations are outlined. Numerical examples are given to illustrate the application of the given equations to the design of laser radar systems.

Laser radars, because of their narrow beamwidths, usually have poor searching capability. These systems, however, are used in conjunction with passive infrared search and tracking systems and thermal imaging systems to obtain target range and reduce false alarms (detections). In these applications IRST or thermal imagers locate the potential targets in a large volume search. Subsequently, the laser radar system is pointed in the direction of the target to obtain range. In this application the laser system can also confirm the existence of the target.

Laser radars can also be used in conjunction with microwave and millimeter wave radars in a similar manner to improve angular position and range accuracy of the targets.

8.1 INTRODUCTION

Laser radar systems are active devices that operate in a manner very similar to microwave radars but at a much higher frequency. This higher frequency has the beneficial effect of smaller components and remarkable angular accuracy values. The atmospheric attenuation losses, however, are considerable at these high frequencies.

Ground-based laser systems usually have a limited detection range (about 10 km), primarily because of atmospheric effects. Spaceborne laser radar systems,

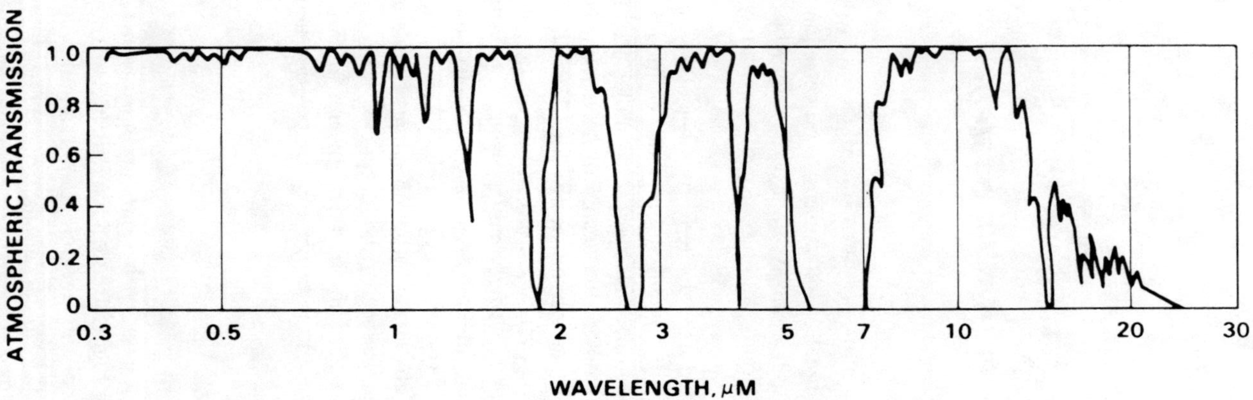

Figure 8.1 Electromagnetic radiation through the atmosphere at μm wavelengths.

however, can be built with detection ranges of thousands of kilometers in the absence of atmospheric attenuation.

Laser radars measure radar-target range and angular positions in much the same way as microwave radars. The small angular beam spread of laser radars (in microradians) severely limits the ability of these radars in large volume search applications. In a number of cases, large volume search devices, such as microwave radars and infrared systems, are used as initial search and target locating devices. This coarse target angular position is subsequently used for queuing of the laser radar systems.

Primarily because of the existence of atmospheric attenuation windows and the availability of components, laser radars are built at specific wavelengths. Figure 8.1 shows atmospheric windows at laser radar wavelengths. From this figure we see transmission coefficients of unity at the 1 and 10 μm regions. Most of the operating laser radars are constructed at 1.06 and 10.6 μm wavelengths. The 1.06 μm lasers are referred to as *Neodymium YAG crystal lasers*, whereas 10.6 μm laser radars are called *CO_2 gas lasers*. The solid-state laser system of 1.06 μm is not as efficient as the CO_2 gas lasers of 10.6 μm. The former may have an efficiency of 1% or 2%, whereas the latter may have an efficiency of about 10%.

8.2 PERFORMANCE CALCULATIONS

Referring to the geometry of Figure 8.2, the signal power received by a laser radar from a target at range R can be written as

$$S = \underset{1}{\frac{P}{4\pi R^2}} \cdot \underset{2}{\frac{4\pi}{\frac{\pi}{4}\theta_{\mathrm{BW}}^2}} \cdot \underset{3}{\frac{\pi}{4}(R\theta_{\mathrm{DW}})^2} \cdot \underset{4}{\rho} \cdot \underset{5}{\frac{1}{2\pi R^2}} \cdot \underset{6}{A} \cdot \underset{7}{\tau_o} \tag{8.1}$$

where

S is the target signal power at the laser radar,
P is the transmitted pulse power,
R is the radar-target range,
θ_{BW} is the laser beamwidth,
ρ is the target backscattering coefficient,
A is the lens aperture area,
τ_o is the optical efficiency.

The first term of eq. (8.1) represents the distribution of the point source power P over a spherical area of $4\pi R^2$. This would be the power distribution of a point source without the benefit of the lens, which focuses the power in a given direction.

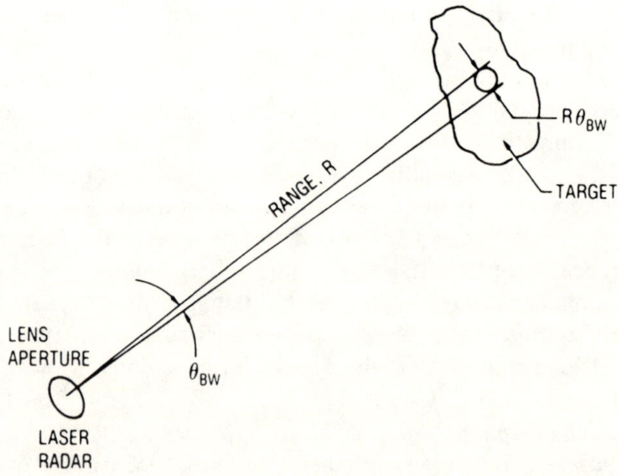

Figure 8.2 Laser radar-target geometry.

The second term is a measure of focusing effect of the lens where it is used to direct the radiated power in a given direction. This gain is the ratio of the spread of power over a sphere of 4π sr and the laser beamwidth, also expressed in steradians.

The third term of eq. (8.1) represents the area of the target. In this derivation, the laser beam is assumed to be considerably smaller than the target area. This assumption can be removed if the target is smaller than the laser beam spread at radar-target range R, as will be shown later.

The fourth term is the backscattering coefficient of the target. This value depends on the surface condition of the target, such as roughness, wetness, and so on. The fifth term is the mathematical expression of the fact that the power at the target is also distributed and spread over a hemispherical area similar to the transmitted power of the first term. In eq. (8.1), the power at the target is assumed to spread over the forward hemisphere of 2π sr. The resulting power density at laser range R will be $1/2\pi R^2$.

Another assumption, often used with respect to the distribution of power at the target, is that of Lambertian scattering. Lambertian distribution assumes that the reflected flux per unit solid angle is proportional to the cosine of the angle between the direction of interest and the plane normal to the surface. Referring to Figure 8.3, we can compute the "steradian" spread of target power as

$$\int_0^{\pi/2} 2\pi \sin\theta \cos\theta \, d\theta = [2\pi(\sin^2\theta/2)]_0^{\pi/2} = \pi \tag{8.2}$$

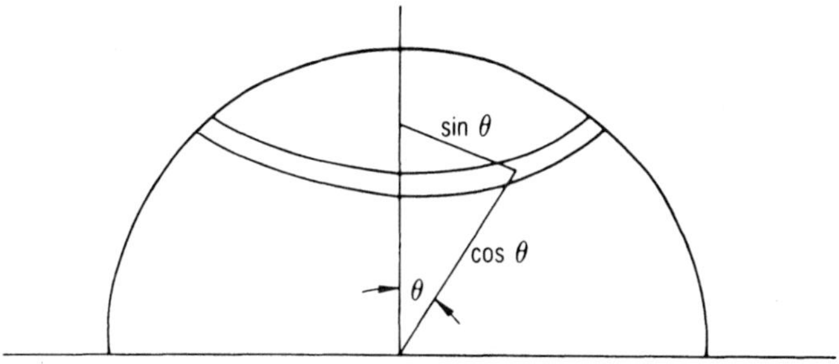

Figure 8.3 Lambertian distribution of incident power at the target.

Therefore, in the case of Lambertian assumption, the fifth term of eq. (8.2) will be divided by π instead of 2π.

Target return power is intercepted by the lens area A as given by the sixth term of eq. (8.1). The seventh term represents the optical efficiency of the laser radar transmission chain following the front aperture of the lens. The target signal power of eq. (8.1) can be simplified to the forms of eqs. (8.3a) and (8.3b) for Lambertian and semispherical target power distribution, respectively:

$$S = \frac{PA\tau_o \rho}{\pi R^2} \tag{8.3a}$$

$$S = \frac{PA\tau_o \rho}{2\pi R^2} \tag{8.3b}$$

According to eq. (8.3), the signal power of the target is inversely proportional to the square of the laser radar-target range R. Note that in the derivation of eq. (8.3), we assumed that the laser beam is smaller than the target. The following analysis will show that where the laser radar beam is larger than the target (point target), the signal power is inversely proportional to the fourth power of the laser-target range. This is the same as in the case of microwave radars.

8.3 LASER BEAM LARGER THAN TARGET

If the target does not intercept the entire laser beam as expressed by the third term of eq. (8.1), this term should be substituted by the laser *radar cross section* (RCS). This target cross section is denoted by σ, equation (8.1) becomes

$$S = \frac{2PA\sigma\rho\tau_o}{\pi^2 R^4 \, \theta_{\mathrm{BW}}^2} \tag{8.4}$$

Equation (8.4) can be described in terms of laser radar parameters by substituting the laser radar beamwidth in terms of lens diameter and wavelength. The expression for beamwidth can be written

$$\theta_{BW} = \frac{\lambda}{D} \tag{8.5}$$

where λ is the wavelength and D is the lens diameter.

At this point, a discussion of beamwidths expressed as a function of lens dimensions is appropriate. In microwave radars, the 3-dB or one-half power beamwidth usually is applied. This value for a $\sin x/x$ beam structure is equal to $0.88\ \lambda/D$, expressed in radians. In optical systems, $1/e = 0.367$ instead of one-half power point is used. This power level will result in a beamwidth of $1.05\ \lambda/D$, again expressed in radians. From these values, note that the usual beamwidth approximation of λ/D can be considered appropriate for both microwave and laser radars.

Using eq. (8.5) and expressing the lens area as

$$A = \frac{\pi}{4}D^2 \tag{8.6}$$

in eq. (8.4), we get

$$S = \frac{8PA^2\sigma\rho\tau_o}{\pi^3 R^4 \lambda^2} \tag{8.7}$$

From eq. (8.7), note that the target signal is inversely proportional to the fourth power of laser radar-target range.

8.4 FLOODLIGHT LASER RADAR SYSTEMS

In a number of applications, two different laser radars are used for transmitting and receiving functions: (1) target illumination, that is transmission of energy towards the target, and (2) receipt of portions of this energy scattered from the target. Figure 8.4 shows the geometry of floodlight lasers. In these lasers, the transmitting beamwidth usually is larger than the receiving beamwidth—hence the name *floodlight lasers*. This larger transmitting beam circumvents the problem of target search.

The target signal power, eq. (8.1), can be adopted easily to floodlight laser radars as follows:

$$S = \frac{P}{4\pi R_1^2} \cdot \frac{4\pi}{\frac{\pi}{4}\theta_{BW1}^2} \cdot \sigma \cdot \rho \cdot \frac{1}{2\pi R_2^2} \cdot A_2 \cdot \tau_o \tag{8.8}$$

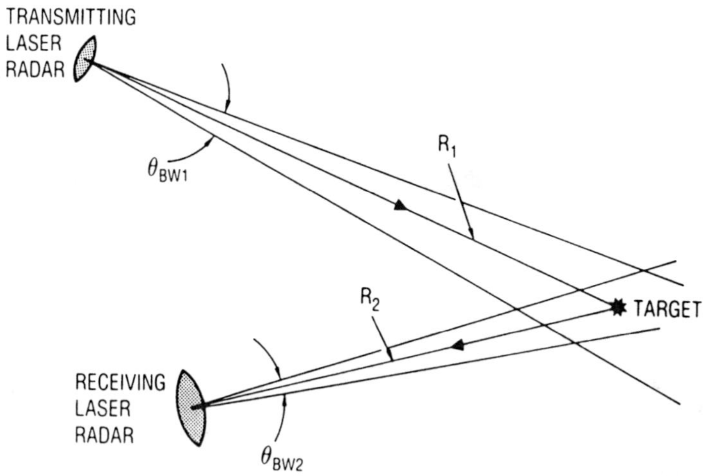

Figure 8.4 Floodlight laser radar transmitting-receiving system.

Where the subscripts 1 and 2 are used for transmitting and receiving functions. In eq. (8.8), we assume that the transmitting beam is considerably larger than the target area at range R_1. The transmitting lens beamwidth in terms of lens area can be written as

$$\begin{aligned} \theta_{BW1}^2 &= \frac{\lambda^2}{D_1^2} \\ &= \frac{\pi\lambda^2}{4A_1} \end{aligned} \tag{8.9}$$

Using this in eq. (8.8) and simplifying, we get

$$S = \frac{8PA_1A_2\sigma\rho\tau_o}{\pi^3\lambda^2R_1^2R_2^2} \tag{8.10}$$

This signal power equation is similar to the *semiactive* microwave radar performance equation. Note the presence of transmitting and receiving apertures and the corresponding laser radar-target ranges in eq. (8.10).

Note also that the laser radar cross section of the target σ appearing in eq. (8.10) is the cross section seen by the receiving laser radar. This cross section usually has a different value than the target cross section, where both the transmitter and the receiver are at the same location.

8.5 LASER RADAR RECEIVERS

Laser radar receivers are of two types: direct and heterodyne detection [1,2]. Direct receivers are similar to electro-optical detectors where the backscattered energy from the target impinges on the detector and is converted to an electrical voltage or an electrical current. Heterodyne receivers, on the other hand, contain a stable local oscillator that is used for frequency reference between transmitting and receiving waveforms.

In heterodyne receivers, the received waveform from the target is heterodyned (multiplied) by the transmitted frequency to reduce the frequency of the returned waveform. At these reduced frequencies, the target signal can be amplified, enhancing the detection process. Note that in heterodyne laser radar, the coherency (phase relationship) of transmitted and received signals is preserved. This allows calculation of the laser radar-target closing rate $\dot{R}$ by doppler processing. Direct detection lasers operate with noncoherent signals, where $\dot{R}$ calculations are not possible.

Figure 8.5 shows the block diagram of direct and heterodyne laser radar receives. In the heterodyne receiver, the transmitted waveform is generated by a *continuous wave* (CW) source. This CW waveform is used as a reference to reduce the frequency of the received waveform from the target.

The detector noise level for *direct* detection is given as

$$N = (A_d \Delta f)^{1/2} / D^* \tag{8.11}$$

where

- N is the noise level, W,
- A_d is the detector area, cm^2,
- Δf is the receiver bandwidth, Hz,
- D^* is the detectivity, cm-$Hz^{1/2}$/W.

The detector area, as seen from Figure 8.6 for a square detector, is related to the lens diameter and focal length as follows:

$$\begin{aligned} A_d &= \alpha_d^{1/2} f_1 \, \alpha_d^{1/2} f_1 \\ &= \alpha_d f_1^2 \end{aligned} \tag{8.12}$$

The focal length f_1 is given as

$$f_1 = (f/\#)D \tag{8.13}$$

where $f/\#$ is the lens focal number. The value of $\alpha_d^{1/2}$ of eq. (8.12) is the instantaneous detector field of view and A_d is the detector area.

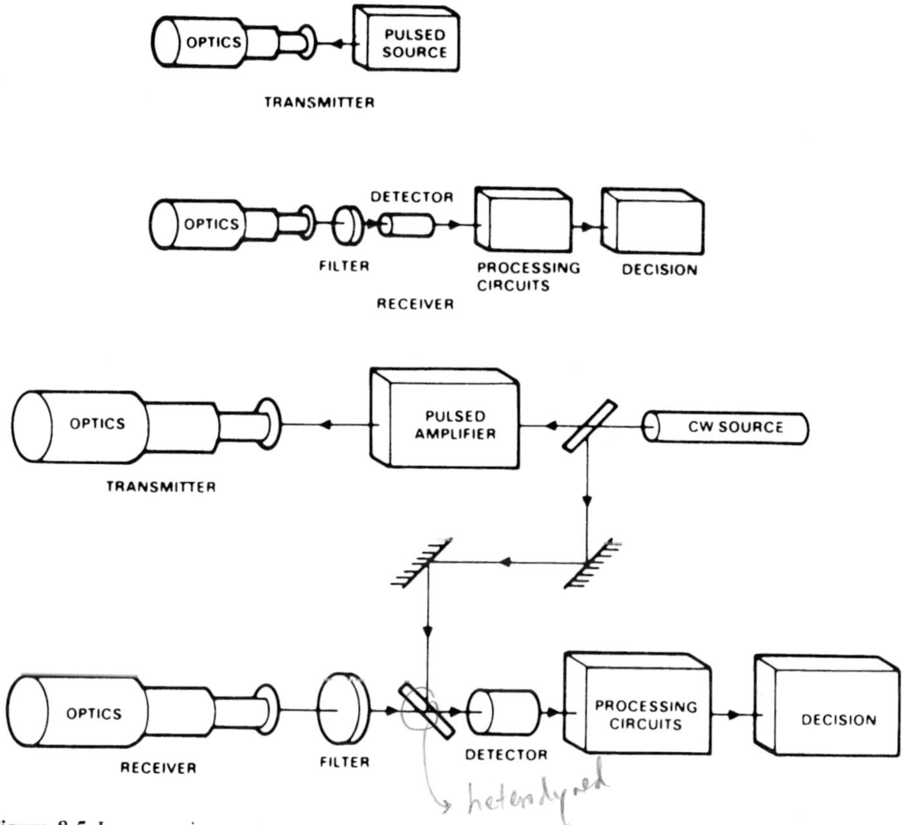

Figure 8.5 Laser receiver system.

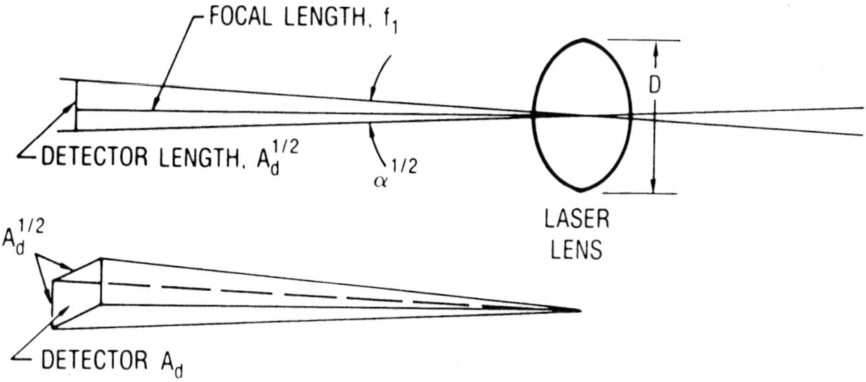

Figure 8.6 Detector area and location with respect to laser radar lens.

The noise level of a *heterodyne* receiver, as given in [3], can be computed as follows. Quantum mechanical analysis has shown that an ideal amplifier has a noise power spectral density of

$$\psi(f) = \frac{hf}{\exp(hf/kT) - 1} + hf \tag{8.14}$$

where

$\psi(f)$ is the spectral density, W/Hz,
h is the Planck's constant, $= 6.6256E - 34$ W-s^2,
f is the frequency, Hz,
k is Boltzmann constant, $= 1.38E - 23$ W-s/K,
T is the absolute temperature, K.

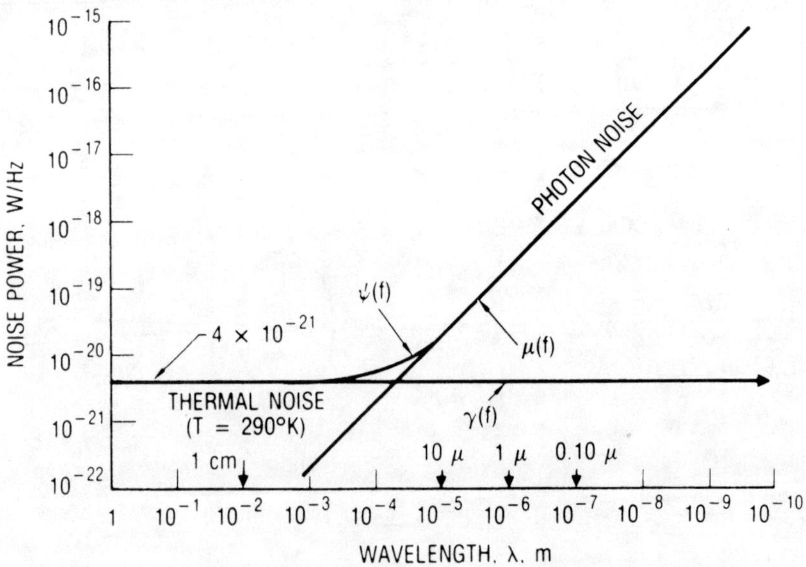

Figure 8.7 Noise level of an ideal amplifier as a function of wavelength.

Equation (8.14) can be evaluated as a function of wavelength, using the relation

$$\lambda = \frac{c}{f} \tag{8.15}$$

where c is the velocity of propagation and f is the frequency. The noise power plot of eq. (8.14) is shown in Figure 8.7. The value of $\psi(f)$ of this figure is that eq. (8.14). This figure also includes curves of

$$\mu(f) = hf \tag{8.16}$$

$$\gamma(f) = kT \tag{8.17}$$

Note that at centimeter wavelengths, with frequencies corresponding to microwave radars, the noise power is given by eq. (8.17). At micrometer wavelengths, the noise power is more correctly approximated by eq. (8.16). Using this, the noise level of a *heterodyne* receiver of a laser radar system can be written as

$$N = (hf/\eta)(B) \tag{8.18}$$

where

η is the quantum efficiency,
B is the receiver bandwidth.

Quantum efficiencies usually are on the order of 30% to 50%.

8.6 PROBABILITY OF DETECTION

The previously given equations can be used to compute the target signal-to-noise ratio of a given laser radar against a specified target. This signal-to-noise ratio can be used to compute the probability of detection with threshold setting as a parameter. Before discussing the probability of detection, we shall describe types of target returns encountered by laser radar systems.

Laser radar targets can be divided into two general categories: speckled and glint targets. These target returns and their statistical behavior are extensively discussed in [4]. In summary, we can categorize targets as having several reflecting points or a single dominating reflector. Targets consisting of a single dominating reflector (such as corner reflectors or normal surfaces) usually give target returns from pulse to pulse that are more or less constant. These targets are referred to as *glint targets*. A histogram plot of glint targets will reveal a Gaussian distribution around the dominating reflector. The standard deviation of this distribution usually is small compared to its average value. Figure 8.8 gives a normalized histogram of a glint target return. In this case, the target consisted of a retroreflector. One hundred pulse returns were used in the construction of this figure. The average value is 1 with a standard deviation of 0.067.

When we have signal-to-noise ratio of a glint target, we can compute the probability of detection of this target using the curves of Figure 8.9. These curves are standard probability of detection curves of a sinusoidal signal in Gaussian noise (see Chapter 6). Figure 8.9 includes the probability of false alarm as a parameter. This parameter is used to set the receiver threshold as a function of desired false alarm rate, as discussed in Chapter 6.

In a precise analysis of the probability of detection of glint targets, a small correction should be made to account for the spread of the target signal around the average value as given in [5]. For most cases, however, this correction is less than 1 dB in signal power.

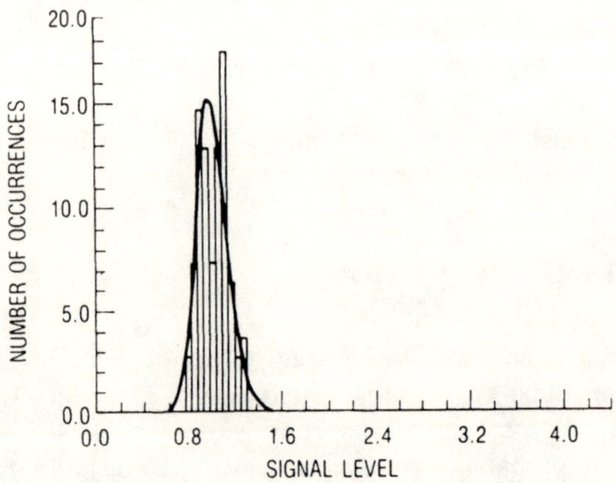

Figure 8.8 Histogram of a glint target signal power, showing a Gaussian probability distribution (average value = 1.0; standard deviation = 0.067).

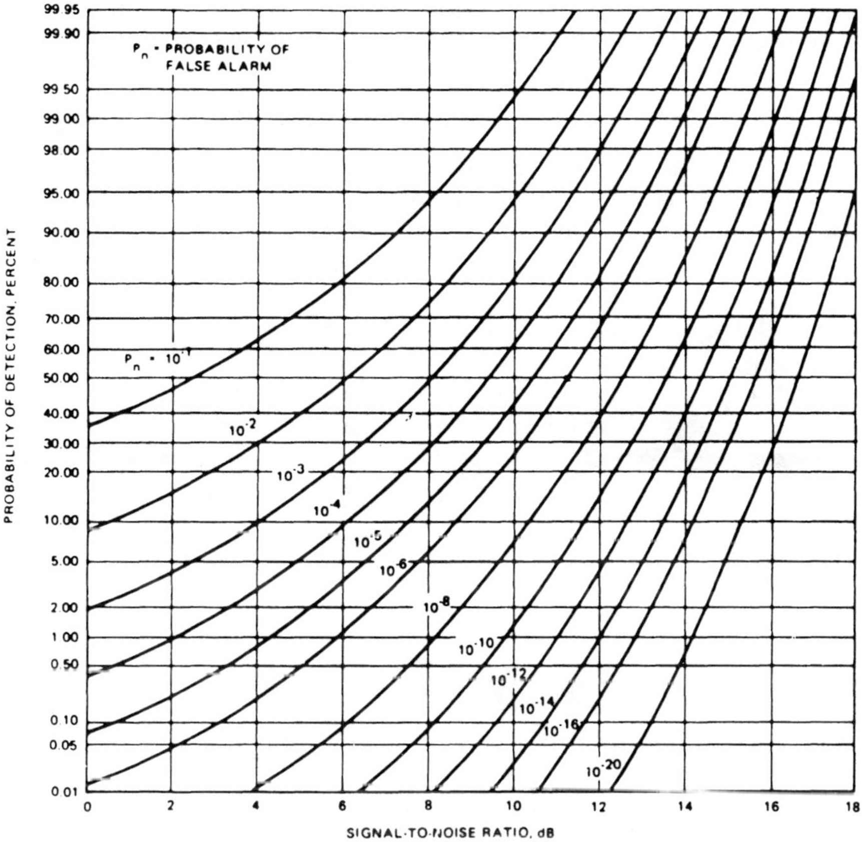

Figure 8.9 Probability of detection as a function of signal-to-noise ratio for glint targets.

8.7 SPECKLED TARGETS

Targets with several reflecting points, each with somewhat equal backscattering characteristics, tend to give speckled power returns. Assume that the power return from each reflecting point behaves with a Gaussian probability distribution. The sum of the scattered target power, representing the total signal power from the target, can then be represented by a Rayleigh distribution.

Figure 8.10 shows a histogram consisting of 1,200 pulse returns obtained from a speckled target simulated by a flame-sprayed, aluminum calibration plate [6]. Speckled targets also result in radar returns approximating exponential distribution. Both Rayleigh and exponential distributions can be used in conjunction with the probability of detection curves of Figure 8.9 to obtain the probability of detection of speckled targets. In an earlier work [5], we showed that a negligible difference in the probability of detection exists, whether Rayleigh or exponential distribution is used in these calculations. Figure 8.11 shows the probability of detection curves as a function of signal-to-noise ratio for a speckled target. The probability of false alarm is used as a parameter. Comparison of these curves with the glint target curves of Figure 8.9 will show that higher levels of signal-to-noise ratio are required for speckled targets to achieve *above* 30% probability of detection values.

Note that the same probability of detection curves as in Figures 8.9 and 8.11 are used in the case of microwave radars. In microwave radars, glint targets are called *nonscintillating targets*, and speckled targets are called *scintillating targets*.

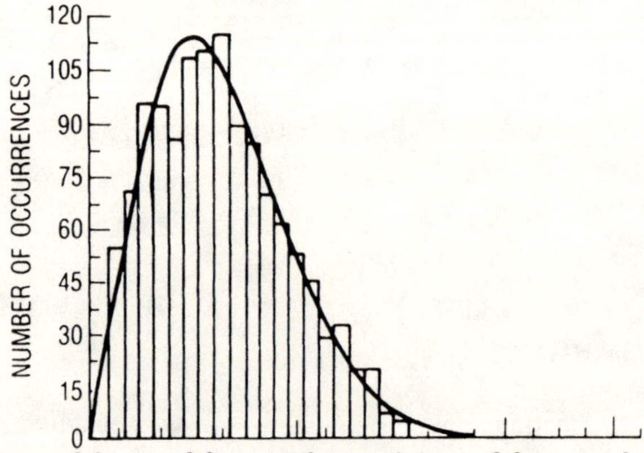

Figure 8.10 Histogram of a speckled laser target power showing a Rayleigh distribution.

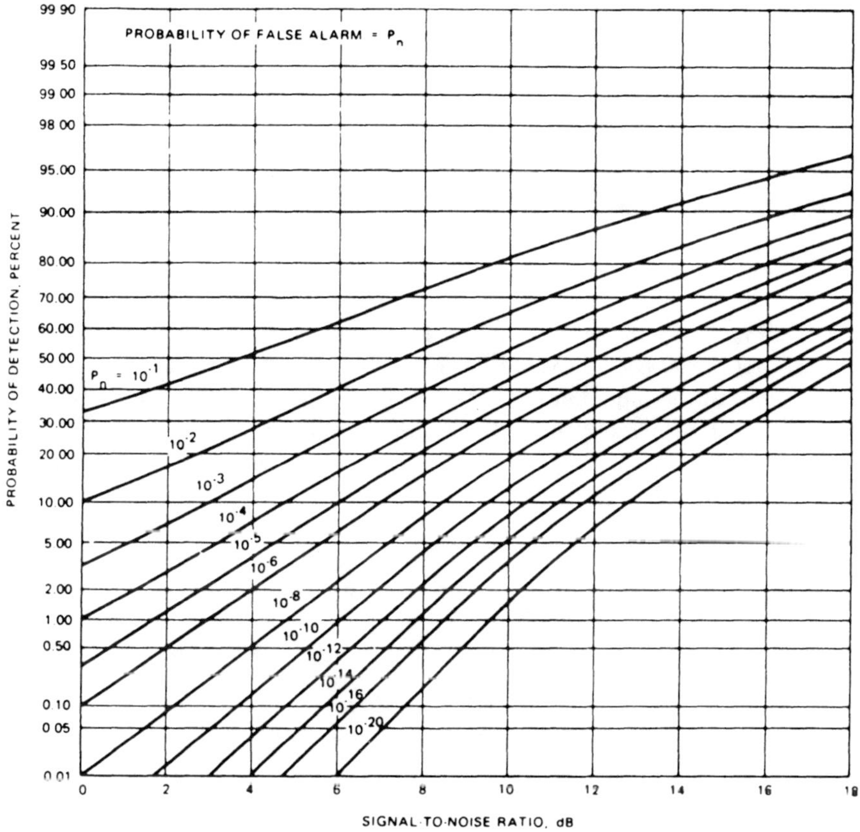

Figure 8.11 Probability of detection versus signal-to-noise ratio for speckled targets with exponential or Rayleigh distribution.

8.8 PROBABILITY OF FALSE ALARMS AND TIME BETWEEN FALSE ALARMS

The previously given probability of detection curves specifies the probability of detection of targets as a function of signal-to-noise ratio with the probability of false alarms as a parameter. The probability of false alarm is used to set the laser radar receiver threshold, as well as to calculate the number of false alarms per unit time or, equivalently, time between false alarms.

The threshold setting for a given false alarm corresponds to 50% probability of detection signal-to-noise ratio of a glint target. For example, from Figure 8.9, the threshold setting for a probability of false alarm of 10^{-6} will correspond to 11.25 dB. This value is an effective power-to-noise ratio; that is,

$$\frac{E_t^2}{2} = 11.25 \text{ dB} \tag{8.19}$$

where E_t is the peak threshold-to-noise voltage ratio. From eq. (8.19), we have

$$E_t = 5.16 \tag{8.20}$$

To compute the number of false alarms per unit time (s), consider the following discussion. Assume a pulsed laser radar system with a pulse duration of τ seconds and range gates of the same duration as shown in Figure 8.12. The number of range gates n examined during 1 second for the presence of targets will be

$$n = \frac{1}{\tau} \tag{8.21}$$

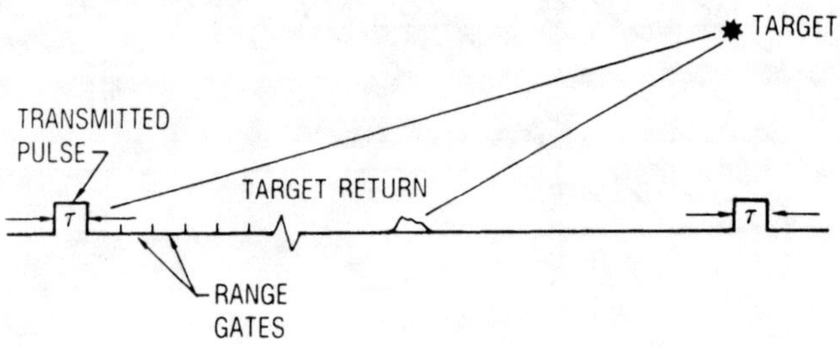

Figure 8.12 Transmitted pulses, range gates, and target returns.

The probability of seeing a false target in each of these range gates is equal to the probability of false alarm P_n. Therefore, the number of false alarms per second N_{fa} can be written

$$N_{fa} = \frac{P_n}{\tau} \tag{8.22}$$

Note that these false alarms are caused only by receiver noise. They are not produced by outside environmental effects, such as clouds, ground emissivities, and weather conditions.

Equation (8.22) gives the number of false alarms in 1 s produced in a random range gate. If we impose the additional condition that a *false alarm* will be considered a *false target detection* if, and only if, it appears in a specified range gate following two consecutive pulse transmissions, we have

$$N_{\mathrm{fa}} = \frac{P_n^2}{\tau} \tag{8.23}$$

This additional restriction, which is easy to implement, will considerably reduce the number of false detections per second.

As an example of eq. (8.23), consider a laser radar with a pulse duration of 1 ns and a probability of false alarm of 10^{-6}. When these values are used, eq. (8.23) results in

$$\begin{aligned} N_{\mathrm{fa}} &= \frac{(10^{-6})^2}{10^{-9}} \\ &= 10^{-3} \end{aligned} \tag{8.24}$$

or one *false target detection* every 16.67 min.

8.9 LASER RADAR MEASUREMENTS

Laser radars are capable of measuring radar-target range, range rate, and angular position much the same way as other microwave radars. The laser radar-target range R is measured by measuring the elapsed time t between transmitting and receiving pulses, to yield

$$R = \frac{ct}{2} \tag{8.25}$$

Heterodyne lasers can measure radar-target range rate R by doppler frequency processing; that is,

$$\Delta f = \frac{2R}{\lambda} \tag{8.26}$$

Where Δf is the measured doppler shift between transmitting and receiving signals, and λ is the wavelength. The angular position of the target is obtained by registering the angular position of the laser beam at the time of target signal return. The target angular position of a laser radar can be further improved by the use of quadrant detector configuration (see Chapter 7). In this mechanization the returned laser pulse impinges on four detectors placed in a quadrant formation. The output of each of four detectors is proportional to the target image power incident on it. Using the resulting detector power levels, a more accurate target angular position location can be computed.

Other range measurement techniques used in microwave radars, such as frequency modulation ranging and multiple *pulse repetition frequency* (PRF) ranging, can be used equally well in laser radars [5]. In fact, [7] gives a detailed discussion of the mechanization procedure of a laser radar using pulse compression methods. In this mechanization, a CO_2 heterodyne laser radar system is used to obtain detection ranges of several kilometers with a ranging accuracy of 10 m and a radial velocity accuracy of less than 1 m/s.

Numerical Example

We wish to design a 10-km ground-based imaging laser radar that satisfies the following requirements:

Coverage	$0.5 \times 0.5°$
Resolution	2×2 ft
Range	10 km

We anticipate that this laser radar will produce images similar to the one given in Figure 8.13. This figure represents a tank-type target. The design of this laser radar can be accomplished through a series of calculations and a series of parameter estimations. These parameters are then used in signal-to-noise ratio and probability of detection calculations.

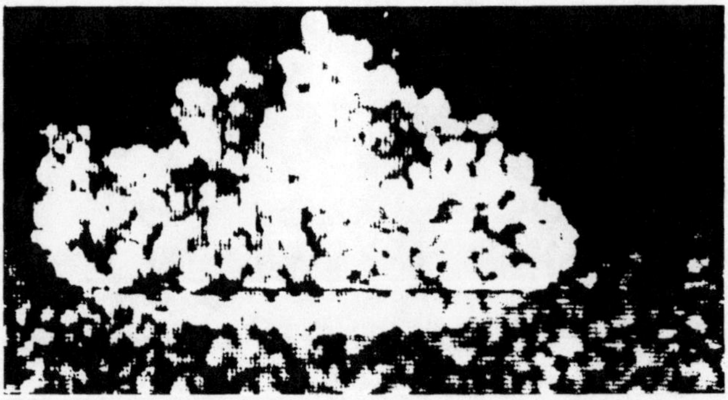

Figure 8.13 Anticipated laser radar image of a tank-type target.

The 2-ft resolution requirement will give rise to a detector instantaneous field of view of

$$\alpha_d^{1/2} = \frac{2}{10{,}000 \times 3.28} = 6.09 \times 10^{-5} \text{ rad} \qquad (8.27)$$

where 10,000 m is the range, and 3.28 is the conversion from meters to feet. Using eq. (8.27), α_d becomes

$$\alpha_d = 3.71 \times 10^{-9} \text{ sr} \qquad (8.28)$$

The coverage of $0.5 \times 0.5°$ can be converted to steradians as follows:

$$\Omega = \frac{0.5 \times 0.5}{(57.3)^2} = 7.6 \times 10^{-5} \text{ sr} \qquad (8.29)$$

Round-trip travel time for 10 km coverage becomes

$$t = \frac{2R}{c} = 6.67\,E - 5 \text{ s} \qquad (8.30)$$

The number of resolution cells or pixels N_p required for the coverage of the target can be computed from eqs. (8.28) and (8.29) as follows:

$$N_p = \frac{\Omega}{\alpha} = 2.048E4 \qquad (8.31)$$

Assuming a search time of 0.1 s for the $0.5 \times 0.5°$ coverage, we find the number of pixels covered per second $\mathring{N}_p$:

$$\mathring{N}_p = \frac{N_p}{0.1} = 2.048E5 \text{ pixels/s} \qquad (8.32)$$

The number of pixels for the round-trip time of 10 km can be computed by using eqs. (8.30) and (8.32) as

$$N' = \mathring{N}_p t$$

$$= 13.66 \tag{8.33}$$

$$= 15$$

This number also represents the number of laser radar receivers required. The reason is that, as the laser radar beam covers the target in the searching mode, target return pulses from previously covered pixels will be arriving. There should be a receiver to receive them. The system can be designed with a transmitter in a TV-type raster scan with 15 receivers following the transmitter. The number of receivers can be reduced by increasing the search time to 1 s or more from its value of 0.1 s.

8.10 SIGNAL-TO-NOISE RATIO COMPUTATION

Consider the design of a CO_2 laser radar for the application of the previous section. The signal-to-noise ratios can be computed from eqs. (8.3a) and (8.18) and written as follows:

$$\frac{S}{N} = \frac{PA\rho\tau_o}{\pi R^2(hf/\eta)(B)} \tag{8.34}$$

where, in eq. (8.34), we assumed that the laser radar beam covers the target, as it should in the imaging laser radar of this design.

Consider the following design parameters for the imaging laser radar:

$$\begin{aligned}
P &= \text{pulse power} \\
&= 20 \text{ W} \\
D &= \text{aperture diameter} \\
&= 8 \text{ in} \\
\rho &= \text{target reflectivity} \\
&= 0.01 \\
\tau_o &= \text{optical transmission} \\
&= 0.40 \\
h &= \text{Planck's constant} \\
&= 6.6256E - 34 \text{ W-s}^2 \\
f &= \text{frequency} \\
&= c/\lambda
\end{aligned}$$

$$\begin{aligned} c &= \text{velocity of propagation} \\ &= 3\,E + 10\ \text{cm/s} \\ \lambda &= \text{wavelength} \\ &= 10.6\ \mu\text{m} \\ \eta &= \text{quantum efficiency} \\ &= 0.50 \\ B &= \text{receiver bandwidth, Hz} \end{aligned}$$

These design parameters can be used directly in the signal-to-noise ratio equation (8.34). It will be more convenient, however, to compute some of these parameters beforehand.

The aperture area A can be computed from lens diameter:

$$\begin{aligned} A &= \frac{\pi}{4} D^2 \\ &= \frac{\pi}{4} (8 \times 2.54 \times 10^{-2})^2 \qquad (8.35) \\ &= 0.0324\ \text{m}^2 \end{aligned}$$

The transmitted frequency f is as follows:

$$\begin{aligned} f &= c/\lambda \\ &= \frac{3E8}{10.6E - 6} \qquad (8.36) \\ &= 2.83\,E13\ \text{Hz} \end{aligned}$$

The receiver bandwidth B can be computed from a consideration of on-target time; that is,

$$B = \frac{1}{\text{OTT}} \qquad (8.37)$$

where OTT is the on-target time. The on-target time can be calculated from eq. (8.32) as follows:

$$\begin{aligned} \text{OTT} &= \frac{1}{\mathring{N}_p} \\ &= 4.88E - 6\ \text{s} \qquad (8.38) \\ &= 5\ \mu\text{s} \end{aligned}$$

This will result in a bandwidth B of

$$B = \frac{1}{5E - 6}$$

$$= 2\,E5 \tag{8.39}$$

$$= 200 \text{ kHz}$$

Note that the IFOV of the detector from eq. (8.27) is about 60 μrad. This value is slightly larger than the diffraction limit of the 8-in lens; that is,

$$\theta_d = \frac{\lambda}{D}$$

$$= \frac{10.6E - 6}{8 \times 2.54 \times E2} \tag{8.40}$$

$$= 52 \ \mu\text{rad}$$

where θ_d is the lens diffraction limit. This condition should be met in an acceptable design.

Using the given parameters in the signal-to-noise ratio equation (8.34), we can compute the S/N as a function of range R:

$$\frac{S}{N} = \frac{(20)(0.0324)(0.01)(0.40)}{\pi R^2(6.6256E - 34)(2.83E13)(1/0.50)(2E5)}$$

$$= \frac{1.10\,E11}{R^2}\ (R \text{ in meters}) \tag{8.41}$$

$$= \frac{1.10\,E5}{R^2}\ (R \text{ in kilometers})$$

These values of S/N can be converted to dB units and plotted as a function of range as shown in Figure 8.14. For example, at a range of 5 km, the S/N will be

$$\frac{S}{N} = \frac{1.10\,E5}{(5)^2}$$

$$= 36.43 \text{ dB} \tag{8.42}$$

The straight line of Figure 8.14 shows the *free space* S/N versus range curve. These values can be modified for various weather condition losses occurring at transmitted frequencies. Approximate atmospheric losses at 10.6 μm wavelength for several nominal conditions are given in Table 8.1. Figure 8.14 also includes the S/N curves which include atmospheric losses computed with values from Table 8.1. Note that at a given range the two-way atmospheric loss is subtracted from the free space S/N ratio.

Because the laser beam covers the target, we may assume a glint target detection condition. From Figure 8.9, a 95% probability of detection, with false alarm probability of 10 $E - 7$, will require 14 dB of S/N ratio. Considering other unforeseen problems, let us require a 20 dB S/N for detection as given by the horizontal line of Figure 8.14. The intersection of this line with various curves of Figure

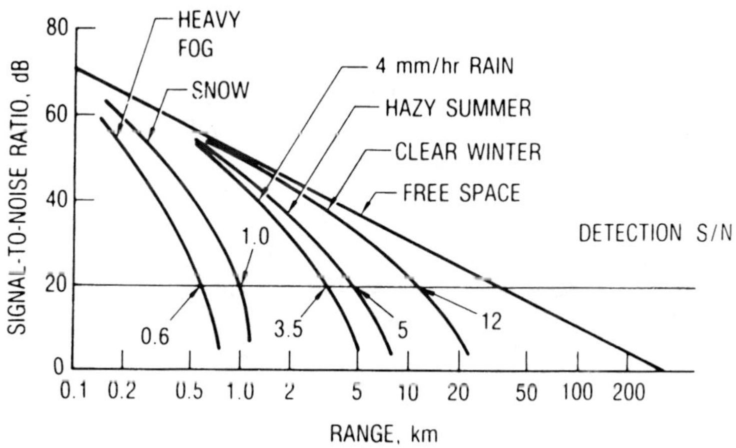

Figure 8.14 CO_2 laser S/N ratio versus range, including atmospheric losses for the imaging laser radar system.

Table 8.1
Atmospheric Attenuation Losses at 10.6 μm (from [3])

Condition	*Visibility, km*	*Loss, dB/km*
Clear winter day	12	0.4
Hazy summer day	5	1.7
Rainfall 4 mm/hr	3.5	3
Snow (1 mm/hr wet)	1	15
Heavy fog	0.6	30

8.14 will give the detection range of the laser radar for the weather conditions under study. These values are summarized in Table 8.2.

Table 8.2
Summary of Detection Ranges of Imaging Laser Radar System

Condition	*Detection Range (km)*
Clear winter	12
Hazy summer	5
Rain (4 mm/hr)	3.5
Snow	1
Heavy fog	0.6

The detection ranges of Table 8.2 happened to be identical to the visibility ranges of Table 8.1; This is not unusual for electro-optical systems such as thermal imagers and laser radars. The difference between TV-type optical systems and thermal imagers lies in the fact that thermal imagers operating on temperature gradients can operate at night as well as during the day. The laser radar systems, having their own source of radiated power, can operate during the day or night without the utilization of temperature gradient or daylight; Laser radar systems, additionally, provide radar-target range and range rate measurements.

From Figure 8.9 the false alarm probability of 10^{-7} will correspond to a threshold setting of 12 dB. This value is the 50% probability of detection signal-to-noise ratio.

Using Eq. (8.22), we find that the number of false alarms per second will be

$$N_{fa} = \frac{P_n}{\text{OTT}} = 0.02 \tag{8.43}$$

or one false alarm every 50 s per detector. If 15 receivers (detectors) are used, then the number of false alarms will be one every 3.3 s. Because this is an imaging laser radar system, the number of false alarms is of little concern.

8.11 GROUND BACKSCATTERING AND INTERFERENCE SIGNALS

In cases where transmitted laser pulses are intercepted by the ground, the signal-to-noise ratio can be computed by using the ground area as the radar target cross section

σ. Referring to Figure 8.15, this area is calculated as follows:

$$\sigma = (R\theta_{BW})\left(\frac{c\tau}{2}\right) \tag{8.44}$$

where σ is the cross section of the laser radar target. The first term after the equal sign is the beam coverage on the ground. The second term represents the equivalent range of a single laser pulse as shown in Figure 8.12. If this pulsewidth is greater than the laser beam coverage, the term should be replaced by the beam coverage on the ground. In this case, eq. (8.44) becomes

$$\sigma = (R\theta_{BW})(R\theta_{BW}/\sin\varepsilon) \tag{8.45}$$

Where ε is the depression angle. Equation (8.45) approximates the elliptical area coverage of the laser beam on the ground. The laser radar cross-section values of eqs. (8.44) and (8.45), together with appropriate reflectivity characteristics ρ, are used in the signal power equations (8.7) and (8.10). The value of the reflectance parameter ρ depends on the intercepted ground terrain (sand, cement, vegetation) as well as laser radar wavelength.

Interference signals, unlike the backscattered energy discussed previously, are caused by natural radiation emittance of the sky, clouds, and ground. This radiation, in some cases, serves to increase the noise level of the laser radar receiver. Natural radiation emittance usually is expressed in

$$\mathrm{W/cm^2 - sr\text{-}\mu m} \tag{8.46}$$

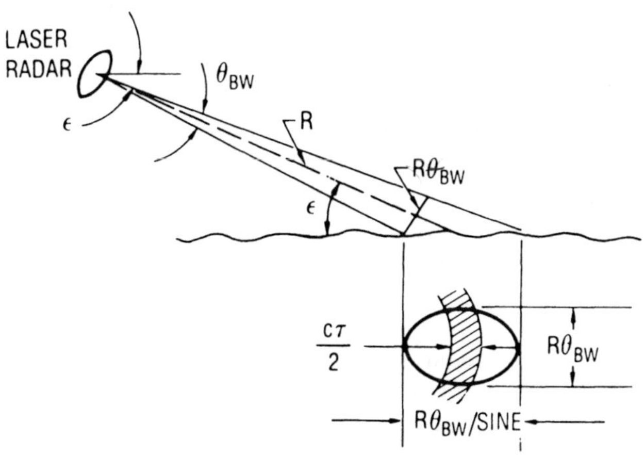

Figure 8.15 Laser radar beam interception with ground.

Therefore, the emitted radiation is per unit area of the radiating surface, with a distribution over a spherical angular coverage in steradians and over a wavelength determined by the laser radar receiver.

For a given laser radar, the background noise power level can be expressed as

$$N_B = \left[\frac{\pi}{4}(R\theta_{\mathrm{BW}})^2\right]\left[\frac{1}{R^2}\right]\left[\frac{\pi D^2}{4}\right] \cdot \int_{\lambda_1}^{\lambda_2} W(\lambda)\, \mathrm{d}\lambda \tag{8.47}$$

where

N_B is the background noise,
D is the lens diameter,
θ_{BW} is the lens beamwidth,
R is the laser-background range,
$W(\lambda)$ is the spectral emittance, $\mathrm{W/cm^2 - sr\text{-}\mu m}$.

The first term of eq. (8.47) represents the area of laser beam at range R. The product of the second and the third terms represents the equivalent lens solid angle in steradians. The integral represents background emission.

Equation (8.47) can be simplified into the following form:

$$N_B = \frac{\pi^2}{16} D^2 \theta_{\mathrm{BW}}^2 \int_{\lambda_1}^{\lambda_2} W(\lambda)\, \mathrm{d}\lambda \tag{8.48}$$

The values of spectral emittance as a function of wavelength are measured and represented in Figures 8.16 and 8.17. Figure 8.16 gives the radiance of a cloudless sky background as a function of wavelength. Figure 8.17 represents solar reflection from clouds and Earth, as well as self-emissions from Earth.

Note from Figure 8.17 that solar reflection decreases exponentially with increasing wavelength. Self-emission from Earth, however, increases with increasing wavelength. Because laser radar systems usually operate in a narrow frequency band, we can approximate the integral of eq. (8.48) by assuming $W(\lambda)$ to be constant over this frequency region; that is,

$$N_B = \frac{\pi^2}{16} D^2 \theta_{\mathrm{BW}}^2 W(\lambda) \Delta\lambda \tag{8.49}$$

Where $\Delta\lambda$ is the wavelength equivalent of the laser receiver bandwidth. This value can be obtained from the definition of a wavelength:

$$\lambda = \frac{c}{f} \tag{8.50}$$

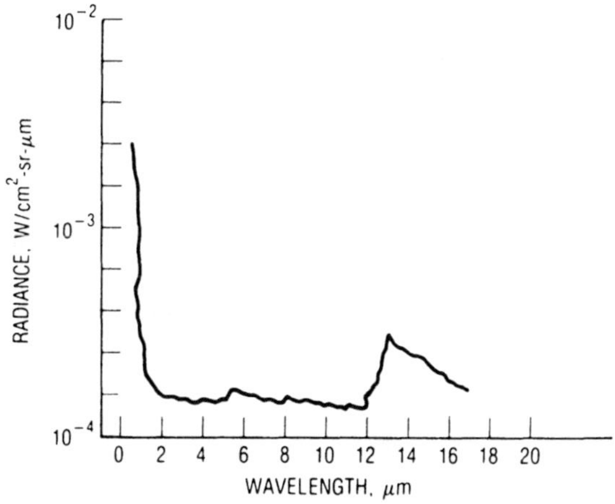

Figure 8.16 Sky background emission for cloudless sky condition.

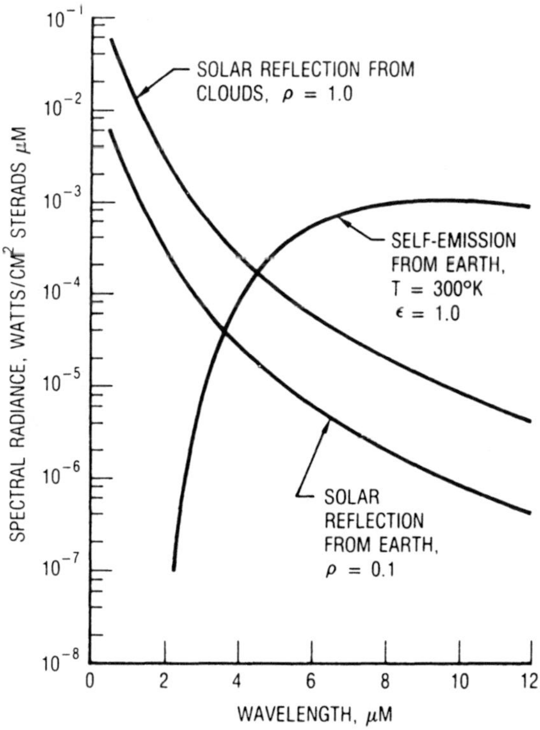

Figure 8.17 Solar reflections from clouds and Earth and self-reflection from Earth.

$$|\Delta\lambda| = cB/f^2$$
$$= \lambda^2 B/c \tag{8.51}$$

where B is the bandwidth of the laser radar receiver. Note that eq. (8.51) gives the "equivalent wavelength" of the receiver bandwidth in a heterodyne receiver. In the case of direct detection receiver the bandwidth $\Delta\lambda$ is determined by optical rejection filter. As a numerical example of background emittance calculation, consider the example of laser radar given above with the following parameters:

$$N = \frac{hf}{\eta} B \tag{8.52}$$

with

$$h = 6.6256\,E - 34 \text{ W-s}^2$$
$$f = 2.83\,E13 \text{ Hz}$$
$$\eta = 0.50$$
$$B = 2E5 \text{ Hz}$$

Using these values in eq. (8.52), we get the noise level of the laser radar receiver

$$N = 7.5\,E - 15 \text{ W}$$
$$= -141 \text{ dB} \tag{8.53}$$

Because this laser radar is designed for ground-level operation, let us compute the noise level caused by Earth's self-emission. For this, we can compute the parameters of eq. (8.49) as follows:

$$D = 8 \text{ in}; \quad \theta_{\text{BW}} = \lambda/D = 52\ \mu\text{rad};$$
$$\Delta\lambda = 7.5\,E - 8\ \mu\text{m}; \quad W(\lambda) = 1\,E - 3 \text{ W/cm}^2\text{-sr-}\mu\text{m} \tag{8.54}$$

where $\Delta\lambda$ is computed from eq. (8.51) and $W(\lambda)$ is obtained from Figure 8.17. Using the parameters of eq. (8.54) in eq. (8.49), we get

$$N_B = 5.16E - 17 \text{ W}$$
$$= -162.8 \text{ dB} \tag{8.55}$$

Note that the above Earth, the self-interference noise level is considerably lower than the system noise level of eq. (8.53). Thus, it is not expected to increase the receiving system noise level.

Note also that $\Delta\lambda$ of eq. (8.51), and the subsequent computed value of eq. (8.54), represent the bandpass equivalent of the laser receiver bandwidth. This is *not* the bandpass of the optical rejection filter, which usually has a bandpass of about 0.15 μm about the transmitted wavelength of 10.6 μm.

8.12 NUMERICAL EXAMPLE OF A SPACEBORNE LASER RADAR SYSTEM

We wish to design a spaceborne NdYAG laser radar system with a target-tracking range of 600 km. After several parameter iterations, we may arrive at the following parameter values:

optics diameter	= 20 cm
wavelength	= 1.06 μm
peak pulse power	= 2 MW
pulsewidth	= 10 ns
target cross section	= 2 m^2
target reflectivity	= 0.1
transmitted PRF	= 100 Hz
tracking S/N	= 3 dB (= 2)
detectivity D^*	= 10 E10 cm-$Hz^{1/2}$/W

Additional laser parameters can be computed or assumed using these data. The laser radar beamwidth can be computed

$$\theta_{BW} = \frac{\lambda}{D} = 5.3\ \mu\text{rad} \tag{8.56}$$

Using an $f/\#$ of 2, the focal length f_1 becomes

$$f_1 = (f/\#)D = 40\ \text{cm} \tag{8.57}$$

When eqs. (8.56) and (8.57) are used in eq. (8.12), the detector area for a matched transmit beam and detector IFOV becomes

$$A_d = f_1^2(\theta_{BW})^2 = 4.5\,E-8\ \text{cm}^2 \tag{8.58}$$

This area is that of a square detector 2 × 2 μm in size. But NdYAG detectors cannot be manufactured at this size; the manufactured detectors are orders of magnitude larger. Assuming a 250 × 250 μm detector, the detector area will be 6.25E − 4 cm^2.

The lens area can be written

$$\begin{aligned} A &= \frac{\pi}{4} D^2 \\ &= 0.0314 \text{ m}^2 \end{aligned} \tag{8.59}$$

The receiver bandwidth can be considered to be the inverse of pulse width as follows:

$$\begin{aligned} B &= \frac{1}{\tau} \\ &= 100\ E6 \text{ Hz} \end{aligned} \tag{8.60}$$

Using the parameters of the example problem in signal equation (8.7), we get

$$\begin{aligned} S &= \frac{8(2\ E6)\ (0.0314)^2\ (2)(0.1)\ (0.5)}{\pi^3 R^4 (1.06\ E - 6)^2} \\ &= 4.5\ E13/R^4 \quad (R \text{ is in meters}) \end{aligned} \tag{8.61}$$

where an optical efficiency τ_o of 0.5 is assumed. The receiver noise can be computed from eq. (8.11) as follows:

$$\begin{aligned} N &= [(6.25\ E - 4)(100\ E6)]^{1/2}/10^{11} \\ &= 2.5\ E - 9 \text{ W} \end{aligned} \tag{8.62}$$

When eqs. (8.61) and (8.62) are used, the signal-to-noise ratio becomes

$$\frac{S}{N} = \frac{1.8\ E22}{R^4} \tag{8.63}$$

Note that S/N ratio of eq. (8.63) is computed on a per pulse basis. Integration of several pulses will increase this S/N proportional to the number of pulses integrated (neglecting integration loss). Assuming that the tracking loop is capable of integrating for 0.25 s and considering the fact that 100 pulses are being transmitted per second, the S/N improvement factor will be 25. Using this and a tracking S/N of 3 dB (factor of 2), we have

$$R^4 = \frac{(1.8E - 22)(25)\text{m}^4}{2} \qquad (8.64)$$
$$R = 688 \text{ km}$$

This tracking range exceeds the original requirement of 600 km.

From the given values, a number of other important laser radar parameters can be computed. The PRF of 100 Hz will correspond to an interpulse period T and an unambiguous range R_a of

$$T = \frac{1}{\text{PRF}} = \frac{2R_a}{c} \qquad (8.65)$$

The unambiguous range is defined as the range from which target pulse return comes within the first interpulse period. From eq. (8.65), we get

$$R_a = 1500 \text{ km} \qquad (8.66)$$

This value exceeds the design range of 600 km, as it should.

The laser radar resolution is limited by the detector size of 250 μm. Using an f_1 of 40 cm, this resolution corresponds to a detector IFOV of

$$\alpha_d^{1/2} = 625 \ \mu\text{rad} \qquad (8.67)$$

which at 600 km corresponds to a linear distance of 375 meters. This resolution validates the assumption of point target. The pulse width τ of 10 ns is equivalent to a resolution range of

$$\Delta R = \frac{ct}{2} = 1.5 \text{ m} \qquad (8.68)$$

These values are within the limits of a target cross section of 2 m^2, which was given in the design of the laser radar.

Background interference can be computed from eq. (8.49) and the data of Figure 8.17. This computation will show that the interference is below the system noise level as given by eq. (8.62).

REFERENCES

[1] Jelalian, A.V. *Laser Radar Systems*, Norwood, MA: Artech House, 1991.

[2] Due, C.T., and L.M. Peterson, *Optical-Mechanical Active/Passive Imaging Systems*, Vol. 1, Ann Arbor: Environmental Research Institute of Michigan, May 1982.

[3] Jelialian, A.V., "Laser Radar Theory and Technology," in *Radar Technology*, E. Brookner (Ed.), Dedham, MA: Artech House, 1977, p. 344.

[4] Shapiro, J.H., R.W. Reinhold, and D. Park, "Performance Analysis of Peak Detecting Laser Radars," *Proc. SPIE*, Vol. 663, 1986, pp. 38–56.

[5] Hovanessian, S.A., *Radar System Design and Analysis*, Dedham, MA: Artech House, 1984.

[6] Papurt, D.M., J.H. Shapiro, and S.T. Lau, "Measured Turbulence and Speckle Effects in Laser Radar Target Return," *SPIE Proc.*, Vol. 415, 1983, pp. 166–178.

[7] Hulme, K.F., et al., "A CO_2 Laser Rangefinder Using Heterodyne Detection and Chirp Pulse Compression," *Optical and Quantum Electronics*, Vol. 13, 1981, pp. 35–45.

[8] Wolfe, W.L., and G.J. Zissis, *The Infrared Handbook*, Ann Arbor: Environmental Research Institute of Michigan, 1978, pp. 22–87.

Index

Printed in the United States
73992LV00004B/15